PREFACE

This book grew out of notes prepared for the courses in the theory of fields that I gave at the University of Kansas during the academic years of 1962–63 and 1964–65. It is intended to be a textbook, rather than a treatise on field theory. Many interesting topics are left untouched; for example, nothing is said concerning transcendental extensions, and only rank-one valuations are treated. Both this omission and this restriction reflect the fact that my courses were intended to prepare the students for the study of algebraic number theory.

Chapter 1 contains the basic results concerning algebraic extensions. Together with sections devoted to separable and inseparable extensions and to normal extensions, there are sections on finite fields, algebraically closed fields, primitive elements, and norms and traces. The reader must be familiar with everything in Chapter 1 before going on to the rest of the book. Chapter 2 is devoted to Galois theory. Besides the fundamental theorem of Galois theory and some examples, it contains discussions of cyclic extensions, Abelian extensions (Kummer theory), and the solutions of polynomial equations by radicals. Chapter 2 concludes with three sections devoted to the study of infinite algebraic extensions. In these sections some topological notions are used, and they may be skipped by the reader not familiar with topology. In fact, knowledge of only the first two sections of Chapter 2 is necessary for reading the remainder of this book.

The study of valuation theory begins with Chapter 3. In this chapter I have placed a number of basic facts concerning valuations. The chapter also contains a thorough discussion of prolongations of valuations. Chapter 4 is concerned with extensions of valuated fields, and in partic-ular, with extensions of complete valuated fields. Chapter 5 contains a proof of the unique factorization theorem for ideals of the ring of integers of an algebraic number field. The treatment is valuation-theoretic throughout. The chapter also contains a discussion of extensions of such fields.

At the end of each chapter are a number of exercises. These range from the very easy to the rather difficult. I have not marked the difficult ones because I believe that the student should try each one; he will discover soon enough which ones are difficult. In addition, it sometimes happens that a student, being unaware that a certain exercise is considered to be difficult by an author (or his teacher), finds a simple solution to that exercise. I cannot stress strongly enough the importance of doing the exercises. Among other reasons for this importance is the fact that the exercises contain many concrete examples of the theoretical results of the text. In several places an exercise is quoted in a proof. That proof cannot be considered as being complete until that exercise has been done.

It would be very difficult to write a book on the theory of fields that does not reflect the influence of Emil Artin. This influence is certainly evident here, especially in the proof of the fundamental theorem of Galois theory and in many sections of Chapters 3 and 4.

I have assumed that the reader of this book has had a first course in abstract algebra, on the level represented by the book *Topics in Algebra* by I. N. Herstein. This course should have included a discussion of linear algebra.

If R is a commutative ring with unity I shall follow the custom of not considering R as a prime ideal of itself. The reader should know the following facts about R: a residue class ring of R modulo a maximal ideal is a field; if R is the ring of polynomials in one indeterminate over a field, then every nonzero prime ideal of R is maximal; the existence and construction of the field of quotients of R if R is an integral domain.

The reader will recall that a field always contains at least two elements, and so the zero element and the unity of a field are not equal.

I believe that the use of Zorn's lemma should be taught in a first course in abstract algebra, and I have used this powerful tool freely in this book.

The symbol $\subseteq$ will be used to indicate containment, while $\subset$ will be reserved for proper containment. When k and K are fields, $k \subseteq K$ always means that k is a subfield of K. When k, L, and K are fields, to say that L is between k and K means that $k \subseteq L$ and $L \subseteq K$. If 1, or 0, is the identity element of a group, I shall use the same symbol for the subgroup of that group consisting of a single element. When k is a field I use k^* to denote both the set of nonzero elements of k and the multiplicative group of k. The symbol $\parallel$ is placed at the end of a proof, and the symbol $\cong$ is used for isomorphism.

The theorems are numbered consecutively throughout each chapter.

Algebraic Extensions of Fields

BY

PAUL J. MCCARTHY
The University of Kansas

DOVER PUBLICATIONS, INC.
New York

DEDICATED TO MY PARENTS

Published in Canada by General Publishing Company, Ltd., 30 Lesmill Road, Don Mills, Toronto, Ontario.
Published in the United Kingdom by Constable and Company, Ltd., 3 The Lanchesters, 162–164 Fulham Palace Road, London W6 9ER.

This Dover edition, first published in 1991, is an unabridged, slightly corrected republication of the Second Edition, published by Chelsea Publishing Company, New York, 1976. The original edition was published by Blaisdell Publishing Company ("A Division of Ginn and Company"), Waltham, Mass., 1966, as "A Blaisdell Book in Pure and Applied Mathematics."

Manufactured in the United States of America
Dover Publications, Inc., 31 East 2nd Street, Mineola, N.Y. 11501

Library of Congress Cataloging-in-Publication Data

McCarthy, Paul J. (Paul Joseph), 1928–
    Algebraic extensions of fields / by Paul J. McCarthy.
        p.   cm.
    "An unabridged, slightly corrected republication of the second edition published by Chelsea Publishing Company, New York, 1976"—T.p. verso.
    Includes bibliographical references and index.
    ISBN 0-486-66651-4
    1. Fields, Algebraic. 2. Field extensions (Mathematics).
I. Title.
QA247.M387   1991
512'.3—dc20   90-49779
                                                             CIP

The numbering of lemmas and propositions begin anew in each section.

Finally, I wish to express my thanks to the following persons: Professors W. R. Scott and George Whaples for reading the manuscript and for making many helpful suggestions; Professor George Springer for urging me to write a book based on my course notes, and for the useful comments he made at the beginning of my efforts; and my teacher, Professor A. E. Ross, for introducing me to the study of field theory and for his encouragement over the years.

P.J.M.

CONTENTS

CHAPTER 1

Algebraic Extensions

1. Definitions

Let k be a field.

DEFINITION. A field K is called an *extension* of k if k is a subfield of K. Let K be an extension of k and let S be a subset of K. Denote by $k(S)$ the smallest subfield of K which contains both k and S. This is evidently the intersection of all the subfields of K which contain both k and S. The field $k(S)$ is an extension of k and we shall say that it is obtained by *adjoining* S to k, or by the *adjunction* of S to k.

Let S and T be subsets of K. Then it is clear that $k(S \cup T) = k(S)(T) = k(T)(S)$. If S is a finite set, say $S = \{a_1, \ldots, a_n\}$, we shall denote $k(S)$ by $k(a_1, \ldots, a_n)$.

If K is an extension of k then it is a linear vector space over k. As such it has a dimension over k, which may be infinite. This dimension is called the *degree* of K over k and is denoted by $[K:k]$.

THEOREM. 1. *If K is an extension of k and L is an extension of K then $[L:k] = [L:K][K:k]$.*

Proof. Let $a_1, \ldots, a_m$ be elements of L that are linearly independent over K and let $b_1, \ldots, b_n$ be elements of K that are linearly independent over k. We shall show that the mn products

$$a_i b_j, \quad i = 1, \ldots, m, \quad j = 1, \ldots, n,$$

are linearly independent over k. Suppose that

$$\sum_{i=1}^{m} \sum_{j=1}^{n} c_{ij} a_i b_j = 0 \quad \text{where each } c_{ij} \in k.$$

1

Then

$$\sum_{i=1}^{m} \left(\sum_{j=1}^{n} c_{ij} b_j \right) a_i = 0,$$

and since for each i, $\sum_{j=1}^{n} c_{ij} b_j \in K$, we have

$$\sum_{j=1}^{n} c_{ij} b_j = 0, \qquad i = 1, \ldots, m.$$

Therefore, $c_{ij} = 0$ for each i and each j.

It follows from what we have just proved that the formula of the theorem holds whenever either $[L : K]$ or $[K : k]$ is infinite. Suppose they are both finite and that $m = [L : K]$ and $n = [K : k]$. Then $a_1, \ldots, a_m$ is a basis of L over K and $b_1, \ldots, b_n$ is a basis of K over k. We shall show that the mn products $a_i b_j$ form a basis of L over k. Let $d \in L$. Then

$$d = \sum_{i=1}^{m} e_i a_i \qquad \text{where each } e_i \in K,$$

and for $i = 1, \ldots, m$,

$$e_i = \sum_{j=1}^{n} f_{ij} b_j \qquad \text{where each } f_{ij} \in k.$$

Hence

$$d = \sum_{i=1}^{m} \sum_{j=1}^{n} f_{ij} a_i b_j,$$

and so the mn products $a_i b_j$ span L as a vector space over k. Since we have already shown that they are linearly independent over k, they form a basis of L over k. ‖

DEFINITION. An extension K of k is called a *finite extension* of k if $[K : k]$ is finite.

If K is a finite extension of k and if $a_1, \ldots, a_n$ is a basis of K over k, then every element of K can be written in the form

$$\sum_{i=1}^{n} b_i a_i, \qquad b_1, \ldots, \qquad b_n \in k.$$

Since these sums are clearly in $k(a_1, \ldots, a_n)$ we have $K = k(a_1, \ldots, a_n)$.

DEFINITION. An extension K of k is called a *simple extension* of k if $K = k(a)$ for some $a \in K$.

It follows that every finite extension of k can be obtained by a succession of simple extensions. If $K = k(a_1, \ldots, a_n)$ and if we define $k_0 = k$ and $k_i = k_{i-1}(a_i)$ for $i = 1, \ldots, n$, then $K = k_n$.

Now let K be an extension of k and assume that $K = k(S)$. Let ω be the cardinal number of the set S. Let $f(x_1, \ldots, x_n)$ be a polynomial in n indeterminates, where $n \leqslant \omega$, with coefficients in k. If $a_1, \ldots, a_n \in S$ then

$f(a_1, \ldots, a_n) \in K$. The set of all such elements of K forms a subring of K which contains k and S, and which we denote by $k[S]$. This ring is certainly an integral domain and therefore has a field of quotients in K. But K itself is the smallest subfield of K which contains both k and S and, consequently, $K = k(S)$ is the field of quotients of $k[S]$ in K. Thus we conclude that $K = k(S)$ consists of all quotients

$$\frac{f(a_1, \ldots, a_n)}{g(b_1, \ldots, b_m)},$$

where $f(x_1, \ldots, x_n)$ and $g(x_1, \ldots, x_m)$ are polynomials in n and m indeterminates, respectively, with coefficients in k, $n \leqslant \omega$, $m \leqslant \omega$, $a_1, \ldots, a_n$, $b_1, \ldots, b_m \in S$, and $g(b_1, \ldots, b_m) \neq 0$.

Consider now the case of a simple extension $K = k(a)$ of k. Denote by $k[x]$ the ring of polynomials in the indeterminate x with coefficients in k: this ring is called the ring of polynomials *over* k. Then $K = k(a)$ is the field of quotients (*in* K) of the integral domain

$$k[a] = \{f(a) \,|\, f(x) \in k[x]\},$$

that is,

$$k(a) = \left\{\frac{f(a)}{g(a)} \,\middle|\, f(x), g(x) \in k[x], g(a) \neq 0\right\}.$$

It is evident that we are led quite naturally to introduce the following concepts which we shall study in detail in the succeeding sections.

DEFINITION. Let K be an extension of k and let $a \in K$. We say that a is *algebraic* over k if there is a nonzero polynomial $f(x) \in k[x]$ such that $f(a) = 0$. If a is not algebraic over k, we say that it is *transcendental* over k. The extension K of k is called an *algebraic extension* of k if each element of K is algebraic over k. On the other hand, if at least one element of K is transcendental over k, then K is called a *transcendental extension* of k.

If we wish to express the fact that "K is an extension of k" we shall usually do so by referring to "the extension K/k." Similarly, we shall say that "K/k is algebraic" when we mean that "K is an algebraic extension of k," etc.

2. Algebraic extensions

THEOREM 2. *If K is a finite extension of k then K is an algebraic extension of k.*

Proof. Let $[K:k] = n$. If $a \in K$ then the $n + 1$ elements $1, a, a^2, \ldots, a^n$ of K are linearly dependent over k. Hence there are elements $c_0, c_1, \ldots,$ $c_n \in k$ such that $c_0 + c_1 a + \cdots + c_n a^n = 0$, and at least one of the c_i is not equal to zero. Then $f(x) = c_0 + c_1 x + \cdots + c_n x^n$ is a nonzero polynomial in $k[x]$ such that $f(a) = 0$. ∥

THEOREM 3. *Let K be an extension of k and let $a \in K$ be algebraic over k. Then there is a unique monic irreduibcle polynomial $p(x) \in k[x]$ such that $p(a) = 0$. If $g(x)$ is any polynomial in $k[x]$ such that $g(a) = 0$, then $p(x)$ divides $g(x)$. Furthermore, $k(a) = k[a]$ and, in fact, every element of $k(a)$ can be written uniquely in the form $r(a)$ where $r(x) \in k[x]$ and either $r(x) = 0$ or $\deg r(x) < \deg p(x)$.*

Proof. Define the mapping ψ from $k[x]$ into $k[a]$ by $\psi(f(x)) = f(a)$. It is clear that ψ is a homomorphism from $k[x]$ onto $k[a]$. Since a is algebraic over k, the kernel of ψ is a nonzero ideal of $k[x]$. Every ideal of $k[x]$ is principal, therefore the kernel of ψ is of the form $(p(x))$ where we may assume that $p(x)$ is monic. Since $k[x]/(p(x))$ is isomorphic to $k[a]$ it is an integral domain. Hence $(p(x))$ is a prime ideal which implies that $p(x)$ is irreducible in $k[x]$: in fact, $p(x)$ is the only monic irreducible polynomial in the ideal $(p(x))$. Since every nonzero prime ideal of $k[x]$ is maximal, $k[x]/(p(x))$ is a field and the same will then be true of $k[a]$. Hence $k(a) = k[a]$, and we have proved all of the theorem except the very last statement. To complete the proof we must show that every residue class $h(x) + (p(x))$ of $(p(x))$ in $k[x]$ contains a unique representative $r(x)$ where either $r(x) = 0$ or $\deg r(x) < \deg p(x)$. However, this follows immediately from the fact that there are *unique* polynomials $q(x), r(x) \in k[x]$ such that $h(x) = q(x)p(x) + r(x)$ and either $r(x) = 0$ or $\deg r(x) < \deg p(x)$. ∥

If $f(x) \in k[x]$ and if a is an element of some extension of k, such that $f(a) = 0$, then a is called a *root* of $f(x)$. The unique monic irreducible polynomial in $k[x]$ having a as a root will be denoted by Irr (k,a).

COROLLARY 1. *Let K be an extension of k and let $a \in K$. If a is algebraic over k then $k(a)/k$ is finite and therefore algebraic. In fact, $[k(a): k] = \deg$ Irr (k,a).*

Proof. Let $n = \deg$ Irr (k,a). Then, by Theorem 3, $1, a, a^2, \ldots, a^{n-1}$ form a basis of $k(a)$ over k. ∥

COROLLARY 2. *Let K be an extension of k and suppose that $a_1, \ldots, a_n \in K$ are algebraic over k. Then $k(a_1, \ldots, a_n)/k$ is finite and therefore algebraic.*

Proof. By induction on n. The corollary is true when $n = 1$ by Corollary 1. Suppose that $n > 1$ and that the corollary is true when n is replaced by $n - 1$. Then $[k(a_1, \ldots, a_{n-1}): k]$ is finite. Since a_n is algebraic over k, and since $k[x] \subseteq k(a_1, \ldots, a_{n-1})[x]$, a_n is algebraic over $k(a_1, \ldots, a_{n-1})$. Hence, by Corollary 1, $[k(a_1, \ldots, a_{n-1}, a_n): k(a_1, \ldots, a_{n-1})]$ is finite. But then, using Theorem 1,

$$[k(a_1, \ldots, a_n): k] = [k(a_1, \ldots, a_n): k(a_1, \ldots, a_{n-1})][k(a_1, \ldots, a_{n-1}): k]$$

is finite. ‖

COROLLARY 3.   *Let K be an extension of k and let a, b ∈ K be algebraic over,
k. Then a + b, a − b, ab, and a/b (when b ≠ 0) are algebraic over k.*

Proof.   By Corollary 2, $k(a, b)$ is algebraic over k, and it contains
$a + b, a − b, ab$, and a/b (when $b \neq 0$). ‖

Let K be an extension of k and let K_a be the set of all those elements of
K which are algebraic over k. If follows from Corollary 3 that K_a is a field
which contains k; it is called the *algebraic closure of k in K*.

DEFINITION.   Let K and L be extensions of k. An isomorphism σ of K
into L is called a *k-isomorphism* if $\sigma(a) = a$ for all $a \in k$. If there is a k-
isomorphism of K *onto* L then K and L are said to be *k-isomorphic*.

DEFINITION.   Let K be an extension of k and let a and b be elements of K
which are algebraic over k. If Irr (k,a) = Irr (k,b) then a and b are said to
be *conjugate* over k, or are called *k-conjugates* of each other.

THEOREM 4.   *Let K be an extension of k and let a and b be elements of K
which are conjugate over k. Then k(a) and k(b) are k-isomorphic. In fact,
there is a k-isomorphism σ of k(a) onto k(b) such that $\sigma(a) = b$.*

Proof.   We define a mapping σ from $k(a)$ onto $k(b)$ as follows: let
$r(x) \in k[x]$ be such that either $r(x) = 0$ or deg $r(x) <$ deg Irr (k,a) and set
$\sigma(r(a)) = r(b)$. It follows immediately from Theorem 3 that σ is a well-
defined one-one mapping from $k(a)$ onto $k(b)$ and that $\sigma(c) = c$ for all
$c \in k$. Let $s(x)$ be another polynomial in $k[x]$ such that either $s(x) = 0$ or
deg $s(x) <$ deg Irr (k,a). Then either $r(x) + s(x) = 0$ or deg $(r(x) + s(x)) <$
deg Irr (k,a) and so $\sigma(r(a) + s(a)) = r(b) + s(b) = \sigma(r(a)) + \sigma(s(a))$. Now
suppose that $r(x)s(x) = q(x)$ Irr $(k,a) + t(x)$ where either $t(x) = 0$ or
deg $t(x) <$ deg Irr (k,a). Then $r(a)s(a) = t(a)$ and $\sigma(r(a)s(a)) = \sigma(t(a)) =$
$t(b) = r(b)s(b) = \sigma(r(a))\sigma(s(a))$. Hence σ is a k-isomorphism. ‖

Up to this point in this section we have assumed that we are given *a priori*
an extension of k which contains elements that are algebraic over k. We
must now question the existence of such extensions. This question is answered by the following important result.

THEOREM 5.   *Let p(x) be a nonconstant irreducible polynomial in k[x]. Then
there is an extension K of k which contains a root of p(x). Furthermore, if
K and L are extensions of k which contain roots a and b, respectively, of p(x),*

then there is a k-isomorphism of the subfield $k(a)$ of K onto the subfield $k(b)$ of L which maps a onto b.

Proof. The residue class ring $K = k[x]/(p(x))$ is a field. We may consider k as a subring of $k[x]$ and if $a, b \in k$, then $a \equiv b(\bmod p(x))$ if and only if $a = b$. Thus, in K, $a + (p(x)) = b + (p(x))$ if and only if $a = b$. Therefore, if we identify $a \in k$ and $a + (p(x)) \in K$, we may consider k as a subfield of K. The element $x + (p(x))$ of K is a root of $p(x)$, for $p(x + (p(x))) = p(x) + (p(x)) = (p(x))$, which is the zero element of K. The last statement of the theorem is proved in exactly the same way as Theorem 4. ‖

We close this section with two examples which illustrate some of the ideas and results that we have been discussing. Here, and throughout this book, the field of rational numbers will be denoted by Q.

Example 1.   Let $p(x) = x^3 - x + 1$. This polynomial is irreducible in $Q[x]$ and, by Theorem 5, there exists a field $Q(a)$ where a is a root of $p(x)$. By Theorem 3 every element of $Q(a)$ can be written uniquely in the form $r(a)$, where $r(x) \in Q[x]$ and either $r(x) = 0$ or $\deg r(x) \leqslant 2$. Let $b = 2 - 3a + 2a^2$. We shall determine b^{-1}. We know that $b^{-1} = c_0 + c_1 a + c_2 a^2$, where c_0, c_1, c_2 are rational numbers, and

$$\begin{aligned} 1 = bb^{-1} &= (2 - 3a + 2a^2)(c_0 + c_1 a + c_2 a^2) \\ &= 2c_0 + (2c_1 - 3c_0)a \\ &\quad + (2c_2 - 3c_1 + 2c_0)a^2 \\ &\quad + (-3c_2 + 2c_1)a^3 + 2c_2 a^4. \end{aligned}$$

Since $p(a) = 0$ we have $a^3 = a - 1$ and $a^4 = a^2 - a$. Therefore,

$$1 = 2c_0 - 2c_1 + 3c_2 + (-3c_0 + 4c_1 - 5c_2)a + (2c_0 - 3c_1 + 4c_2)a^2.$$

Now 1 can be expressed in only one way as a linear combination of 1, a, and a^2 with rational coefficients and so we must have

$$\begin{aligned} 2c_0 - 2c_1 + 3c_2 &= 1 \\ -3c_0 + 4c_1 - 5c_2 &= 0 \\ 2c_0 - 3c_1 + 4c_2 &= 0. \end{aligned}$$

If we solve this system of equations we obtain $c_0 = 1$, $c_1 = 2$, and $c_2 = 1$, so that $b^{-1} = 1 + 2a + a^2$.

Example 2.   Consider again the field $Q(a)$ where a is a root of $p(x) = x^3 - x + 1$. Since $Q(a)$ is a finite extension of Q it is algebraic over Q.

Let $b = 2 - 3a + 2a^2$. Then b is algebraic over Q and we can determine Irr (Q,b). We have

$$b = 2 - 3a + 2a^2$$
$$ab = -2 + 4a - 3a^2$$
$$a^2b = 3 - 5a + 4a^2.$$

We can rewrite these equations as

$$(2 - b) - 3a + 2a^2 = 0,$$
$$-2 + (4 - b)a - 3a^2 = 0,$$
$$3 - 5a + (4 - b)a^2 = 0.$$

Here we have a system of three homogeneous equations with coefficients in $Q(a)$, which has a nontrivial solution in $Q(a)$, namely, 1, a, a^2. Hence the determinant of coefficients must be zero. Thus b is a root of the polynomial

$$f(x) = - \begin{vmatrix} 2 - x & -3 & 2 \\ -2 & 4 - x & -3 \\ 3 & -5 & 4 - x \end{vmatrix}$$
$$= x^3 - 10x^2 + 5x - 1.$$

This polynomial is irreducible in $Q[x]$: thus Irr $(Q,b) = f(x)$.

3. Characteristic: perfect fields

In a field k the intersection of all of the subfields of k is a subfield of k, which is clearly the smallest subfield of k. It is called the *prime field* of k and will be denoted by Δ in the following discussion. We shall determine Δ, at least up to isomorphism.

Let J be the ring of integers and let e be the identity element of k (usually we denote this element by 1, but it is more convenient to use e in this section). If $n \in J$ we define

$$ne = \begin{cases} e + e + \cdots + e \, (n \, e\text{'s}) & \text{if } n > 0, \\ 0 & \text{if } n = 0, \\ -(-n)e & \text{if } n < 0. \end{cases}$$

If we define the mapping ϕ from J into k by $\phi(n) = ne$ for $n \in J$, then ϕ is a homomorphism from J into k. Actually it maps J into Δ since $e \in \Delta$. The kernel of this homomorphism cannot be all of J since $\phi(1) = e \neq 0$. Hence we have only the following two cases.

Case 1. Suppose that ϕ is an isomorphism. Then Δ has a subring which is isomorphic to J and therefore must have a subfield isomorphic to the field of quotients of J, namely, the field Q of rational numbers. However, Δ must be this subfield, and consequently $\Delta \cong Q$.

Case 2. Suppose that the kernel of ϕ is an ideal of J different from the zero ideal. This ideal is a principal ideal different from J and, since J modulo this ideal is an integral domain (it is isomorphic to a subring of the field Δ), it must be generated by a prime p. Thus Δ has a subring isomorphic to $J/(p)$. But $J/(p)$ is a field and so $\Delta \cong J/(p)$. We have proved

THEOREM 6.   *The prime field of a field is isomorphic to either Q or $J/(p)$ for some prime p.*

DEFINITION.   If the prime field of a field k is isomorphic to $J/(p)$ for the prime p, then we say that k has *characteristic p*. Otherwise we say that k has *characteristic zero*. The characteristic of k will be abbreviated char k.

Let $a \in k$ and let n be an integer. Set $na = (ne)a$. If char $k = 0$, then $na = 0$ if and only if either $n = 0$ or $a = 0$. If char $k = p$ and if $a \neq 0$, then $na = 0$ if and only if p divides n.

Let K be an extension of k and let a be an element of K which is algebraic over k. Let $f(x) \in k[x]$ be such that $f(a) = 0$ and assume that $f(x) = (x - a)^m g(x)$ where $g(x) \in K[x]$ and $g(a) \neq 0$. Then a is said to be a *root of multiplicity m* of $f(x)$. If a is a root of multiplicity 1 of $f(x)$ then it is called a *simple root* of $f(x)$.

There is a very convenient criterion for a root of a polynomial to be a simple root that makes use of the concept of the *derivative* of a polynomial. We are able to define the derivative of a polynomial over an arbitrary field without making use of a limiting process. If $f(x) = a_0 + a_1x + a_2x^2 + \cdots + a_nx^n$, we define its derivative to be the polynomial $f'(x) = a_1 + 2a_2x + \cdots + na_nx^{n-1}$. It is easy to show that if $f(x)$ and $g(x)$ are polynomials over the same field, then

$$(f(x) + g(x))' = f'(x) + g'(x) \text{ and}$$
$$(f(x)g(x))' = f(x)g'(x) + f'(x)g(x).$$

PROPOSITION.   *Let K be an extension of k, a an element of K which is algebraic over k, and $f(x)$ a nonzero polynomial in $k[x]$ which has a as a root. Then a is a simple root of $f(x)$ if and only if $f'(a) \neq 0$.*

Proof. If $f(x) = (x - a)g(x)$, where $g(a) \neq 0$, then $f'(x) = (x - a)g'(x) + g(x)$ and as a result $f'(a) = g(a) \neq 0$. Conversely, suppose that $f'(a) \neq 0$ and that $f(x) = (x - a)^m g(x)$ where $g(a) \neq 0$. If $m > 1$, then $f'(x) = m(x - a)^{m-1}g(x) + (x - a)^m g'(x)$, and since $m - 1 > 0$ we have $f'(a) = 0$, a contradiction. Thus we must have $m = 1$. ‖

DEFINITION.   Let K be an extension of k. An element a of K which is algebraic over k is said to be *separable* over k if it is a simple root of Irr (k,a).

The extension K is said to be a *separable extension* of k if it is algebraic over k and if each of its elements is separable over k. We also say that K is *separable over k* or that K/k is *separable*. If K/k is algebraic but not separable, we say that K is an *inseparable extension* of k.

DEFINITION. A field k is called *perfect* if it has no inseparable extensions.

Suppose that char $k = p$ and consider the mapping from k into itself given by $a \rightarrow a^p$. It is clear that $(ab)^p = a^p b^p$, and we also have $(a + b)^p = a^p + b^p$. For,

$$(a + b)^p = a^p + b^p + \sum_{n=1}^{p-1} \binom{p}{n} a^n b^{p-n}.$$

If $1 \leqslant n \leqslant p - 1$, the binomial coefficient $\binom{p}{n}$ is divisible by p and all the terms in the summation are zero. Thus the mapping in question is a homomorphism of the field k into itself. Since every homomorphism of one field into another is either an isomorphism or trivial, and since $e^p = e$, this mapping must be an isomorphism of k into itself. The set of images under this isomorphism is a subfield of k: this subfield is

$$k^p = \{a^p \mid a \in k\}.$$

Note that if $a \in k$ then a has at most one pth root in any extension of k. We have $a \in k^p$ if and only if a has a pth root in k.

THEOREM 7. *The field k is perfect if and only if either* char $k = 0$ *or* char $k = p$ *and* $k^p = k$.

Proof. Suppose char $k = 0$. Let K be an algebraic extension of k, $a \in K$, and $p(x) = \text{Irr}(k, a)$. Then $p'(x) \neq 0$ and $p'(x)$ is of smaller degree than $p(x)$. Hence $p(x)$ cannot divide $p'(x)$; it follows therefore from Theorem 3 that $p'(a) \neq 0$. Thus a is separable over k and we conclude that K/k is separable.

Now suppose that char $k = p$ and let K, a, and $p(x)$ be as above. Assume that $k^p = k$ and that a is not separable over k. Since $p'(a) = 0$ and since $\deg p'(x) < \deg p(x)$, it follows from Theorem 3 that $p'(x) = 0$. Let bx^m be a typical term of $p(x)$. Then $mbx^{m-1} = 0$ and we must have either $b = 0$ or m divisible by p. Thus

$$p(x) = \sum_{r=0}^{n} a_r x^{pr}.$$

Since $k^p = k$ each a_r has a unique pth root in k. Let $b_r^p = a_r$ for $r = 0, 1, \ldots, n$, then

$$p(x) = \sum_{r=0}^{n} b_r^p x^{pr} = \left(\sum_{r=0}^{n} b_r x^r \right)^p,$$

which contradicts the fact that $p(x)$ is irreducible in $k[x]$. Thus K/k must be separable and it follows that k is perfect.

Conversely, suppose that char $k = p$ and that k is perfect. Let $a \in k$ and consider the polynomial $x^p - a \in k[x]$. If this polynomial has a root b in k then $a = b^p \in k^p$. Suppose it has no root in k and let $p(x)$ be one of its nonconstant monic irreducible factors in $k[x]$. Consider the extension $k(b)$ where $p(b) = 0$. In $k(b)[x]$ we have $x^p - a = x^p - b^p = (x - b)^p$ and since $p(x)$ divides this polynomial we have $p(x) = (x - b)^m$ for some m. If $m = 1$ then $x - b \in k[x]$ and so $b \in k$ which is not true. Hence $m > 1$. But then b is not a simple root of $p(x)$ and since $p(x) = \mathrm{Irr}(k, b)$, $k(b)$ is not separable over k. This contradicts the fact that k is perfect. Therefore we must have $a \in k^p$ for all $a \in k$, that is, $k^p = k$. ∥

COROLLARY. *Every finite field is perfect.*

4. Separability of extensions

In this section k will be a field of characteristic p and K a finite, and so algebraic extension of k.

DEFINITION. Let $a \in K$ and let $p(x) = \mathrm{Irr}(k, a)$. Then a is said to be *purely inseparable* over k if $p(x) = (x - a)^m$ for some m.

THEOREM 8. *Let $a \in K$ be purely inseparable over k. Then there is an integer $e \geqslant 0$ such that $\deg \mathrm{Irr}(k, a) = p^e$. Then $a^{p^e} \in k$ but $a^{p^f} \notin k$ for $0 \leqslant f < e$.*

Proof. Let $\deg \mathrm{Irr}(k, a) = mp^e$ where m is not divisible by p. If $p(x) = \mathrm{Irr}(k, a)$ then $p(x) = (x - a)^{mp^e} = (x^{p^e} - a^{p^e})^m$. Since each coefficient of $p(x)$ lies in k we have $ma^{p^e} \in k$. Since p does not divide m we conclude that $a^{p^e} \in k$. Then, since $p(x)$ is irreducible in $k[x]$, we must have $m = 1$. If $a^{p^f} \in k$ for some non-negative $f < e$ and if $f(x) = x^{p^f} - a^{p^f}$ then $f(x) \in k[x]$ and $f(a) = 0$. This implies that $p(x)$ divides $f(x)$, which is impossible. ∥

DEFINITION. An extension K of k is said to be *purely inseparable* over k if it is a finite extension of k and if each of its elements is purely inseparable over k.

COROLLARY. *If K is a purely inseparable extension of k then $[K : k]$ is a power of p.*

Proof. Since K is a finite extension of k it is obtained by a number of successive simple extensions

$$k_1 = k(a_1), \qquad k_2 = k_1(a_2), \ldots, \qquad K = k_n = k_{n-1}(a_1).$$

The result will follow from Theorem 8 and Theorem 1 if we show that for $m = 2, \ldots, n$, a_m is purely inseparable over k_{m-1}. Let $p(x) = \mathrm{Irr}(k, a_m)$

and $q(x) = \mathrm{Irr}(k_{m-1}, a_m)$. Then $q(x)$ divides $p(x)$ and since $p(x)$ is a power of $x - a_m$ the same is true of $q(x)$. ||

We shall now prove a theorem which has a corollary that is the converse of the first statement of Theorem 8.

THEOREM 9. *Let a be an element of k which is not in k^p. Then for every integer $e \geqslant 0$ the polynomial $f(x) = x^{p^e} - a$ is irreducible in $k[x]$.*

Proof. Let $p(x)$ be a nonconstant monic irreducible factor of $f(x)$ in $k[x]$ and let $K = k(b)$ where $p(b) = 0$. Then $f(b) = 0$. Thus $a = b^{p^e}$ and so, in $K[x]$, we have $f(x) = (x - b)^{p^e}$. If $q(x)$ is any nonconstant monic irreducible factor of $f(x)$ in $k[x]$, then $q(x)$ is a power of $x - b$ and so $q(b) = 0$. Hence $p(x)$ divides $q(x)$ by Theorem 3, and therefore $q(x) = p(x)$. This means that $f(x)$ is a power of $p(x)$, say $f(x) = p(x)^m$. Since $\deg f(x) = p^e$ both $\deg p(x)$ and m must be powers of p, so that $f(x) = p(x)^{p^r}$ for some integer $r \geqslant 0$. Let c be the constant term of $p(x)$ and recall that $c \in k$ since $p(x) \in k[x]$. Then $a = (\pm c)^{p^r}$. If $r > 0$ this implies that $a \in k^p$ which is contrary to the hypothesis of the theorem. Thus $r = 0$ and $f(x) = p(x)$ is irreducible in $k[x]$. ||

COROLLARY 1. *If $a \in K$ and if there is an integer $e \geqslant 0$ such that $a^{p^e} \in k$ then a is purely inseparable over k.*

Proof. Let e be the smallest non-negative integer such that $a^{p^e} \in k$. If $e = 0$ then $a \in k$ and so it is certainly purely inseparable over k. Hence we assume that $e > 0$. If $a^{p^e} \in k^p$, say $a^{p^e} = b^p$, then $a^{p^{e-1}} = b \in k$ which is contrary to our choice of e. Thus $a^{p^e} \notin k^p$ and it follows from Theorem 9 that $x^{p^e} - a^{p^e}$ is irreducible in $k[x]$. However, this means that $\mathrm{Irr}(k, a) = (x - a)^{p^e}$ which tells us that a is purely inseparable over k. ||

COROLLARY 2. *If a is an element of K which is both separable and purely inseparable over k then $a \in k$.*

Proof. Let e be the smallest non-negative integer such that $a^{p^e} \in k$ and let $a^{p^e} = b$. By Theorem 9, $\mathrm{Irr}(k, a) = x^{p^e} - b$. But a is separable over k and therefore the derivative of this polynomial must not be zero. Hence $e = 0$ and $a \in k$. ||

Let L be an arbitrary field and let M and N be subfields of L. By the *composite* of M and N (in L) we mean the intersection of all the subfields of of L which contain both M and N. This is the smallest subfield of L which contains both M and N and is denoted by MN. Clearly, we have $MN = M(N) = N(M)$. If F is a subfield of both M and N and if $M = F(S)$ for some subset S of L, then $MN = N(M) = N(F(S)) = N(F \cup S) = N(S)$. Using the concept of composite we obtain an important criterion for the separability of K over k.

THEOREM 10. *If K is a separable extension of k and if $p =$ char k then $kK^p = K$: this is true even if $[K: k]$ is infinite. If K/k is finite and $kK^p = K$, then K/k is separable.*

Proof. We shall first assume that K/k is separable. Since $K^p \subseteq kK^p$ we have by Corollary 1 to Theorem 9 that each element of K is purely inseparable over kK^p. However, $k \subseteq kK^p \subseteq K$ and K/k is separable. From this it follows that K/kK^p is separable (see Exercise 18). Hence, by Corollary 2 to Theorem 9, $K \subseteq kK^p$. Therefore $K = kK^p$.

Now assume that K/k is finite and that $K = kK^p$. Let $a \in K$ and let $p(x) = $ Irr (k,a). Suppose that a is not separable over k. Then $p'(a) = 0$ and so $p(x)$ divides $p'(x)$. Arguing as we have done before, we conclude that

$$p(x) = \sum_{r=0}^{m} a_r x^{rp}$$

for some m, where $a_0, a_1, \ldots, a_m \in k$ and $a_m = 1$. Furthermore, not all of the a_i are zero and since $p(a) = 0$ it follows that $1, a^p, a^{2p}, \ldots, a^{mp}$ are linearly dependent over k. However, $1, a, a^2, \ldots, a^m$ must be linearly independent over k, for if they were linearly dependent over k, we could find a nonzero polynomial $f(x) \in k[x]$ with $f(a) = 0$, but $\deg f(x) \leqslant m < \deg p(x)$. This is impossible by Theorem 3.

To arrive at the desired contradiction we shall show that if $b_1, \ldots, b_m$ are elements of K which are linearly independent over k then $b_1{}^p, \ldots, b_m{}^p$ are linearly independent over k. We may assume that $b_1, \ldots, b_m$ form a basis of K/k. If $c \in K$ then $c = d_1 b_1 + \cdots + d_m b_m$ where $d_1, \ldots, d_m \in k$. Then $c^p = d_1{}^p b_1{}^p + \cdots + d_m{}^p b_m{}^p$ so that $b_1{}^p, \ldots, b_m{}^p$ span K^p over k^p, that is,

$$K^p = k^p b_1{}^p + \cdots + k^p b_m{}^p.$$

Then $K = kK^p = kk^p b_1{}^p + \cdots + kk^p b_m{}^p = kb_1{}^p + \cdots + kb_m{}^p$, and so $b_1{}^p, \ldots, b_m{}^p$ span K over k. Now every set of vectors which spans a vector space over a field contains a basis of the vector space over the field. But every basis of K over k has m elements and so $b_1{}^p, \ldots, b_m{}^p$ is a basis of K over k. As such, these elements of K are linearly independent over k. ‖

COROLLARY 1. *If a is an element of K which is separable over k then $k(a) = k(a^p)$ and $k(a)/k$ is separable. If $k(a) = k(a^p)$ then a is separable over k.*

Proof. Suppose a is separable over k. Then a is both separable and purely inseparable over $k(a^p)$. Hence $a \in k(a^p)$ and so $k(a) = k(a^p)$. From this it follows that $kk(a)^p = kk^p(a^p) = k(a^p) = k(a)$, and so $k(a)$ is separable over k. ‖

COROLLARY 2. *If L is a finite separable extension of k and K is a finite separable extension of L then K is separable over k.*

Proof. By hypothesis we have $kL^p = L$ and $LK^p = K$. Hence $K = LK^p = kL^pK^p \subseteq kK^p$ and so $K = kK^p$. Thus K/k is separable. ‖

COROLLARY 3. *If $a_1, \ldots, a_n$ are elements of K which are separable over k then $k(a_1, \ldots, a_n)$ is separable over k.*

Proof. Let

$$k_1 = k(a_1), \ldots, \qquad k_n = k(a_1, \ldots, a_n) = k_{n-1}(a_n).$$

By Corollary 1, k_1/k is separable. Suppose that $i > 1$ and that we have shown that k_{i-1}/k is separable. Since a_i is separable over k, it is separable over k_{i-1}. Then, by Corollary 1, k_i/k_{i-1} is separable and so k_i/k is separable by Corollary 2. When $i = n$ we obtain the desired result. ‖

Remark. The result of Corollary 2 remains true even when the degrees of the extensions are infinite because separability is what one might call a "local" property. Suppose that k, L, and K are as in Corollary 2 but that no assumption is made about the finiteness of the degrees of these extensions. If $a \in K$ let $b_0, b_1, \ldots, b_r$ be the coefficients of Irr (L,a). Let $L' = k(b_0, b_1, \ldots, b_r)$. By Corollary 3, L' is separable over k. Furthermore, Irr $(L',a) = $ Irr (L,a) and so a is separable over L'. By Corollary 1, $L'(a)/L'$ is separable and so by Corollary 2, $L'(a)/k$ is separable. Thus a is separable over k. Since a is an arbitrary element of K it follows that K/k is separable.

COROLLARY 4. *If K is an arbitrary extension of k then*

$$K_s = \{a \in K \mid a \text{ is separable over } k\}$$

is a subfield of K and $k \subseteq K_s$.

This final corollary is an immediate consequence of all that we have done before. The field K_s is called the *separable closure* of k in K.

Let K be an arbitrary extension of k, a an element of K which is algebraic over k, and $p(x) = $ Irr (k,a). If a is not separable over k then $p'(x) = 0$ and so $p(x) \in k[x^p]$, that is, $p(x) = p_1(x^p)$ where $p_1(x) \in k[x]$. The polynomial $p_1(x)$ is certainly irreducible in $k[x]$ and, in fact, $p_1(x) = $ Irr (k,a^p). Either a^p is separable over k or $p_1(x) = p_2(x^p)$ where $p_2(x) \in k[x]$. in which case $p(x) = p_2(x^{p^2})$. We can continue in this manner until we arrive at a polynomial $p_e(x) \in k[x]$ such that $p(x) = p_e(x^{p^e})$, $p_e(x) = $ Irr (k,a^{p^e}), and $p_e(x) \notin k[x^p]$. Then a^{p^e} is separable over k and e is the smallest such non-negative integer for which this is true. Thus, if a is an element of K which is algebraic over k then there is a non-negative integer e such that $a^{p^e} \in K_s$. Therefore we have the following important result:

THEOREM 11. *Let K be an arbitrary algebraic extension of k. Then K can be obtained by a separable extension followed by a purely inseparable extension.*

If K/k is algebraic then K/K_s is purely inseparable, by the discussion preceding the theorem. The degree $[K_s: k]$ is called the *degree of separability* of K/k and is denoted by $[K: k]_s$. The degree $[K: K_s]$ is called the *degree of inseparability* of K/k and is denoted by $[K: k]_i$. If $[K: k]_i$ is finite, it follows from the corollary to Theorem 8 that it is a power of p, the characteristic of k. Therefore we have

THEOREM 12. *If K is a finite extension of k and if $[K: k]$ is not divisible by p then K/k is separable.*

Let $f(x)$ be an irreducible polynomial of degree n in $k[x]$. Let the non-negative integer e be determined by $f(x) \in k[x^{p^e}]$ but $f(x) \notin k[x^{p^{e+1}}]$. Let $n = n_0 p^e$. Then n_0 is called the *reduced degree* of $f(x)$ and p^e the *degree of inseparability* of $f(x)$.

THEOREM 13. *Let $f(x)$ be an irreducible polynomial in $k[x]$ of reduced degree n_0 and degree of inseparability p^e. Let $k(a)$ be an extension of k obtained by adjoining a root of $f(x)$ to k. Then*

$$[k(a): k]_s = n_0, \qquad [k(a): k]_i = p^e.$$

Proof. By the definition of the degree of inseparability of $f(x)$ we know that a^{p^e} is separable over k and deg Irr $(k, a^{p^e}) = n_0$. Hence $[k(a^{p^e}): k] = n_0$ and $[k(a): k(a^{p^e})] = p^e$. Since $k(a^{p^e})/k$ is separable we have $k(a^{p^e}) \subseteq k(a)_s$. Let $b \in k(a)_s$. Then b is separable over k and so it is separable over $k(a^{p^e})$ (see Exercise 18). However, b is also purely inseparable over $k(a^{p^e})$ (see Exercise 23). Hence, by Corollary 2 to Theorem 9, $b \in k(a^{p^e})$. Thus we have $k(a^{p^e}) = k(a)_s$ from which it follows that $[k(a): k]_s = n_0$. Since $[k(a): k] = [k(a): k]_s [k(a): k]_i = n_0 [k(a): k]_i$ and $[k(a): k] = \deg f(x) = n_0 p^e$ we have $[k(a): k]_i = p^e$. $\|$

Remark. The results of this section hold also when char $k = 0$, in so far as they make sense. In this case, we set $[K: k]_s = [K: k]$ and $[K: k]_i = 1$.

5. Normal extensions

If K is an extension of k and if a polynomial $f(x) \in k[x]$ can be factored into linear factors in $K[x]$ then we say that $f(x)$ *splits* in $K[x]$. Thus, if $f(x)$ splits in $K[x]$ we have

$$f(x) = c(x - a_1) \cdots (x - a_n),$$

where $a_1, \ldots, a_n \in K$. We wish to determine the multiplicity of the roots of $f(x)$, that is, how many times each linear factor occurs in this factorization of $f(x)$. It is sufficient to do this under the additional assumption that $f(x)$ is monic and irreducible in $k[x]$.

THEOREM 14. *Let $f(x)$ be a monic irreducible polynomial in $k[x]$ and let K be an extension of k such that $f(x)$ splits in $K[x]$. If char $k = 0$ then each linear factor of $f(x)$ in $K[x]$ occurs exactly one time. If char $k = p$ and if p^e is the degree of inseparability of $f(x)$ then each linear factor of $f(x)$ in $K[x]$ occurs exactly p^e times.*

Proof. If char $k = 0$ the result follows from the fact that k has no inseparable extensions. Suppose that char $k = p$. We have $f(x) \in k[x^{p^e}]$ but $f(x) \notin k[x^{p^{e+1}}]$. Let $f(x) = g(x^{p^e})$ where $g(x) \in k[x]$. Let $x - a$ be one of the linear factors of $f(x)$ in $K[x]$. Then $g(a^{p^e}) = 0$ and since $g(x) \notin k[x^p]$, a^{p^e} is a simple root of $g(x)$. Let $g(x) = (x - a^{p^e})h(x)$ where $h(a^{p^e}) \neq 0$. Then $f(x) = (x^{p^e} - a^{p^e})h(x^{p^e}) = (x - a)^{p^e}h_1(x)$ where $h_1(x) = h(x^{p^e})$ and so $h_1(a) \neq 0$. ‖

DEFINITION. Let $f(x) \in k[x]$. An extension K of k is called a *splitting field* of $f(x)$ over k if $f(x)$ splits in $K[x]$, that is, if

$$f(x) = c(x - a_1) \cdots (x - a_n),$$

where $a_1, \ldots, a_n \in K$, and if $K = k(a_1, \ldots, a_n)$.

THEOREM 15. *Every nonconstant polynomial $f(x) \in k[x]$ has a splitting field over k and any two splitting fields of $f(x)$ over k are k-isomorphic.*

Proof. To prove the existence of a splitting field of $f(x)$ over k it is sufficient to show that there is an extension K of k such that $f(x)$ splits in $K[x]$. Then the subfield of K obtained by adjoining to k the roots of $f(x)$ in K will be the desired splitting field.

Let $n = \deg f(x)$: we shall prove that the field K exists by induction on n. If $n = 1$ we take $K = k$. Suppose that $n > 1$ and assume that the result is true for all polynomials of degree less than n, that is, assume that if $g(x)$ is a polynomial of degree less than n with coefficients in some field L, then there is an extension F of L such that $g(x)$ splits in $F[x]$. Suppose first that $f(x)$ is reducible in $k[x]$ so that $f(x) = g(x)h(x)$ where $g(x)$, $h(x) \in k[x]$ and $\deg g(x) < n$ and $\deg h(x) < n$. Then there is an extension L of k such that $g(x)$ splits in $L[x]$. Since $h(x) \in L[x]$ there is an extension K of L such that $h(x)$ splits in $K[x]$. Then $f(x)$ splits in $K[x]$. Now suppose that $f(x)$ is irreducible in $k[x]$. By Theorem 5 there is an extension L of k which contains a root a of $f(x)$. In $L[x]$ we have $f(x) = (x - a)g(x)$ where $\deg g(x) = n - 1$. By our induction assumption there is an extension K of L such that $g(x)$ splits in $K[x]$. Then $f(x)$ splits in $K[x]$.

Let K and L be splitting fields of $f(x)$ over k. Let $p(x)$ be a nonconstant irreducible factor of $f(x)$ in $k[x]$. Then K contains a root a_1 of $p(x)$ and L contains a root b_1 of $p(x)$. By Theorem 5 there is a k-isomorphism σ of $k(a_1)$ onto $k(b_1)$ such that $\sigma(a_1) = b_1$. Suppose that

$$f(x) = (x - a_1) \cdots (x - a_r)g(x)$$

in $k(a_1)[x]$ where $g(x)$ has no roots in $k(a_1)$. Then $\sigma(a_1), \ldots, \sigma(a_r)$ are roots of $f(x)$ in $k(b_1)$. Let them be $b_1, \ldots, b_r$. In $k(b_1)[x]$ we have $f(x) = (x - b_1) \cdots (x - b_r)(\sigma g)(x)$ where $(\sigma g)(x)$ is that polynomial in $k(b_1)[x]$ obtained by replacing each coefficient in $g(x)$ by its image under σ. If $(\sigma g)(x)$ has a root c in $k(b_1)$ then $\sigma^{-1}(c)$ is a root of $g(x)$ in $k(a_1)$. Hence $(\sigma g)(x)$ has no roots in $k(b_1)$.

If $g(x)$ is a constant then $K = k(a_1)$ and $L = k(b_1)$ and we are finished. Otherwise, $g(x)$ has a nonconstant irreducible factor $q(x)$ in $k(a_1)[x]$. Then $(\sigma q)(x)$ is a nonconstant irreducible factor of $(\sigma g)(x)$ in $k(b_1)[x]$. Let a_{r+1} be a root of $q(x)$ in K and b_{r+1} a root of $(\sigma q)(x)$ in L. Then there is an isomorphism τ of $k(a_1, a_{r+1})$ onto $k(b_1, b_{r+1})$ such that $\tau(a_{r+1}) = b_{r+1}$ and $\tau(a) = \sigma(a)$ for all $a \in k(a_1)$ (see Exercise 8). Now let $K = k(a_1, \ldots, a_n)$ where $a_1, \ldots, a_n$ are all of the roots of $f(x)$ in K. Continuing as above we obtain a k-isomorphism ρ of K into L. Then if $b_i = \rho(a_i)$ for $i = 1, \ldots, n$, $b_1, \ldots, b_n$ are all of the roots of $f(x)$ in L. Hence $L = k(b_1, \ldots, b_n)$ and so ρ is from K onto L. ‖

DEFINITION. An extension K of k is said to be a *normal extension* of k if it is algebraic over k and if every irreducible polynomial in $k[x]$ which has one root in K splits in $K[x]$. We also say that K is normal over k and that K/k is normal.

THEOREM 16.   *Let K be a splitting field over k of a polynomial $f(x) \in k[x]$. Then K is a normal extension of k.*

Proof.   Let $a_1, \ldots, a_n$ be the roots of $f(x)$ in K so that $K = k(a_1, \ldots, a_n)$. Let $g(x)$ be an irreducible polynomial in $k[x]$ which has one root, say b, in K. Let L be a splitting field of $g(x)$ over K and let b' be any root of $g(x)$ in L. By Theorem 5 there is a k-isomorphism σ of $k(b)$ onto $k(b')$ such that $\sigma(b) = b'$. Under the mapping σ, $f(x)$ is mapped onto itself. It is clear that K is a splitting field of $f(x)$ over $k(b)$. Furthermore, $K(b') = k(a_1, \ldots, a_n, b')$ is a splitting field of $f(x)$ over $k(b')$. By a slight alteration of the proof of Theorem 15 we see that there is an isomorphism τ of K onto $K(b')$ such that $\tau(a) = \sigma(a)$ for all $a \in k(b)$. In particular, $\tau(b) = b'$. Also, $\tau(a_1), \ldots, \tau(a_n)$ are the roots of $f(x)$ in $K(b')$ so they are just $a_1, \ldots, a_n$ in some order. Now $b = h(a_1, \ldots, a_n)$ where $h(x_1, \ldots, x_n)$ is some polynomial in n indeterminates having coefficients in k. Then $b' = \tau(h(a_1, \ldots, a_n)) = h(\tau(a_1), \ldots, \tau(a_n)) \in K$. Thus $g(x)$ splits in $K[x]$. ‖

THEOREM 16a.   *Let K be an algebraic extension of k such that every element of K belongs to some subfield of K which contains k and which is a splitting field over k of some polynomial in $k[x]$. Then K is a normal extension of k.*

Proof.   Let $f(x)$ be an irreducible polynomial in $k[x]$ which has one root, say a, in K. Then a belongs to a subfield L of K such that $k \subseteq L$ and L is a splitting field over k of some polynomial in $k[x]$. By Theorem 16, L/k is normal and so $f(x)$ splits in $L[x]$. Hence $f(x)$ splits in $K[x]$. ‖

THEOREM 17. *Let K be a finite normal extension of k. Then K is the splitting field over k of some polynomial in $k[x]$.*

Proof. Let $K = k(a_1, \ldots, a_n)$ and let $f_i(x) = \text{Irr}\,(k,a_i)$. Since K/k is normal, each $f_i(x)$ splits in $K[x]$. Hence $f(x) = f_1(x) \cdots f_n(x)$ splits in $K[x]$, and K is obtained by adjoining the roots of $f(x)$ to k. Therefore K is a splitting field of $f(x)$ over k. ‖

We have characterized the finite normal extensions of k as the splitting fields over k of polynomials in $k[x]$. Our interest in such extensions arises from the special properties that they possess with regard to their k-automorphisms and the k-isomorphisms between their subfields which contain k. We shall now obtain several of these properties.

THEOREM 18. *Let K be a finite normal extension of k and let F and L be k-isomorphic fields between k and K. Then every k-isomorphism of F onto L can be extended to a k-automorphism of K.*

Proof. By Theorem 17, K is a splitting field over k of some polynomial $f(x)$ in $k[x]$. Then $f(x) \in F[x]$ and K is a splitting field of $f(x)$ over F. Likewise, K is a splitting field of $f(x)$ over L. If σ is a k-isomorphism of F onto L, then a slight alteration of the proof of Theorem 15 shows us that σ can be extended to a k-automorphism of K. ‖

Let K be a finite normal extension of k and let $G(K/k)$ be the set of all k-automorphisms of K. Under the operation of composition of mappings, $G(K/k)$ becomes a group. This is a finite group. To see this we use the fact that K is a splitting field over k of some polynomial in $k[x]$, that is, $K = k(a_1, \ldots, a_n)$ where $a_1, \ldots, a_n$ are all of the roots of $f(x)$ in K. If $\sigma \in G(K/k)$ then $\sigma(a_1), \ldots, \sigma(a_n)$ is a permutation of $a_1, \ldots, a_n$. Clearly, the k-automorphism σ is determined completely by its action on $a_1, \ldots, a_n$ so that there are at most $n!$ elements of $G(K/k)$. We shall now set about to determine the exact order of $G(K/k)$.

THEOREM 19. *Let K be a finite extension of k. Then there is a finite normal extension F of k which contains K and which is the smallest such extension in the sense that if L is a normal extension of k which contains K then there is a K-isomorphism of F into L.*

Proof. Let $K = k(a_1, \ldots, a_n)$ and let $f_i(x) = \text{Irr}\,(k,a_i)$ and $f(x) = f_1(x) \cdots f_n(x)$. Then $f(x) \in K[x]$ and we let F be a splitting field of $f(x)$ over K. Let F_1 be the subfield of F obtained by adjoining all the roots of $f(x)$ in F to k. Then $K \subseteq F_1$ and so $F_1 = F$. Thus F is a splitting field of $f(x)$ over k and, by Theorem 16, F/k is finite normal. Now let L be a normal extension of k which contains K. For $i = 1, \ldots, n, f_i(x)$ has one root in L, namely a_i, and so splits in $L[x]$. Hence $f(x)$ splits in $L[x]$. Then L contains a splitting field of $f(x)$ over K which, by Theorem 15, is K-isomorphic to F. ‖

THEOREM 20. *Let K be a finite extension of k and let L be a normal extension of k such that $k \subseteq K \subseteq L$. Let n_0 be the degree of separability of K/k. Then there are exactly n_0 distinct k-isomorphisms of K onto subfields of L.*

Proof. By Theorem 19 it suffices to prove the theorem when L/k is finite (prove this). We shall prove the theorem by induction on n_0. Suppose $n_0 = 1$: then K/k is purely inseparable. Let σ be a k-isomorphism of K into L and let $a \in K$. If char $k = 0$ then $K = k$ and a is left fixed by σ. If char $k = p$ then there is a non-negative integer e such that $a^{p^e} \in k$. Then $(\sigma(a))^{p^e} = \sigma(a^{p^e}) = a^{p^e}$ and so $\sigma(a) = a$. Hence there is only one k-isomorphism of K into L.

Next, we prove the theorem for the case where K is a simple extension of k, say, $K = k(a)$. By Theorems 13 and 14, a has exactly n_0 conjugates in L, say $a_1 = a, a_2, \ldots, a_{n_0}$. If σ is a k-isomorphism of K into L, then the image of K under σ is $k(a_i)$ for some i, and σ is completely determined by the fact that $\sigma(a) = a_i$. Thus there are exactly n_0 k-isomorphisms of k into L.

Assume that $n_0 > 1$ and that the theorem is true whenever $[K: k]_s < n_0$. Since $n_0 > 1$, there are elements of K which do not belong to k and which are separable over k. Let a be one of these elements. Then $k \subset k(a) \subseteq K_s \subseteq K \subseteq L$. Then L is normal over $k(a)$ (see Exercise 27) and $[K: k(a)]_s = r < n_0$. Hence, by our induction assumption, there are exactly r $k(a)$-isomorphisms $\tau_1, \ldots, \tau_r$ of K into L. Since $k(a)/k$ is separable there are exactly $s = [k(a): k]$ k-isomorphisms $\sigma_1, \ldots, \sigma_s$ of $k(a)$ into L. We have $n_0 = [K_s: k] = [K_s: k(a)][k(a): k] = rs$. By Theorem 18 each σ_i can be extended to a k-automorphism of L. This might be done in several different ways: we can choose one such extension once and for all and denote it also by σ_i. Now consider the n_0 k-isomorphisms of K into L given by $\rho_{ij} = \sigma_i \tau_j, i = 1, \ldots, s, j = 1, \ldots, r$. Let η be any k-isomorphism of K into L. The restriction of η to $k(a)$ is, say, σ_i. Then $\sigma_i^{-1}\eta$ leaves each element of $k(a)$ fixed so that $\sigma_i^{-1}\eta = \tau_j$ for some j. Then $\eta = \sigma_i \tau_j = \rho_{ij}$. Our proof will be complete if we show that the ρ_{ij} are distinct. Suppose that $\rho_{ij} = \rho_{hl}$. Then $\rho_{ij}\tau_j^{-1} = \sigma_i$ and $\rho_{hl}\tau_l^{-1} = \sigma_h$ act similarly on each element of $k(a)$ so that $\sigma_i = \sigma_h$. Then we must have $\tau_j = \tau_l$. ‖

Remark. In general, let k, K, F, and L be field such that $k \subseteq K \subseteq L$ and $k \subseteq F \subseteq L$. If there is a k-isomorphism from K onto F we say that K and F are *k-conjugates* in L. Then we conclude from Theorem 20 that if L/k is normal, K has at most n_0 distinct k-conjugates in L.

If G is any group, we denote the order of G by $\#G$. If we take $K = L$ in Theorem 20 we obtain the

COROLLARY. *If K is a finite normal extension of k then $\#G(K/k) = [K:k]_s$.*

THEOREM 21. *Let K be a finite normal extension of k. Let $a \in K$ and suppose that a is left fixed by each element of $G(K/k)$. Then a is purely inseparable over k.*

Proof. We must show that a is the only root of $p(x) = \text{Irr}(k,a)$ in K: since $p(x)$ splits in $K[x]$, this will imply that a is purely inseparable over k. Let b be another root of $p(x)$ in K. By Theorem 5 there is a k-isomorphism of $k(a)$ onto $k(b)$ which maps a onto b. By Theorem 18 this k-isomorphism can be extended to a k-automorphism of K. By hypothesis a is left fixed by this k-automorphism of K and so $b = a$. ‖

Let K be a finite normal extension of k and let $F = \{a \in K \mid \sigma(a) = a$ for all $\sigma \in G(K/k)\}$. It is easy to show that F is a field and we certainly have $k \subseteq F$. By Theorem 21, F/k is purely inseparable. Therefore we have the following result which may rightly be considered as a corollary to Theorem 21.

COROLLARY. *Let K be a finite normal extension of k. Then K can be obtained by a purely inseparable extension followed by a separable extension.*

6. Finite fields

Let K be a finite field. Since the prime field of a field of characteristic zero has infinitely many elements, K must be of prime characteristic, say, char $k = p$. We may assume that the prime field of K is actually the field of residue classes of integers modulo p, which we shall denote by $GF(p)$. Since K has only a finite number of elements it is certainly a finite extension of $GF(p)$. Let $[K: GF(p)] = n$ and let $a_1, \ldots, a_n$ be a basis of K over $GF(p)$. Then K consists of all expressions of the form

$$c_1 a_1 + \cdots + c_n a_n, \qquad c_1, \ldots, c_n \in GF(p).$$

Thus K has exactly p^n elements.

Let $b_1, \ldots, b_{p^n}$ be all of the elements of K and let $h = p^n - 1$. The multiplicative group of K is a finite group of order h and so $b_i^h = 1$ for all $i = 1, \ldots, p^n$ such that $b_i \neq 0$. Then $b_i^{p^n} = b_i$ for all i. The polynomial $x^{p^n} - x \in GF(p)[x]$ has at most p^n roots in K and, on the other hand, it has p^n roots in K. Hence it splits in $K[x]$ and K is the smallest subfield of K which contains all of these roots, that is, K is a splitting field over its prime field of $x^{p^n} - x$. Therefore we have proved

THEOREM 22. *If K is a finite field of characteristic p, and if K has degree n over its prime field, then K has p^n elements. Any two fields having p^n elements are isomorphic: if they both contain $GF(p)$ then they are $GF(p)$-isomorphic.*

We denote a typical field having p^n elements by $GF(p^n)$. These fields are called *Galois fields* in honor of the French mathematician Galois.

THEOREM 23. *The multiplicative group of a finite field is cyclic.*

Proof. Let the field be $GF(p^n)$ and set $p^n - 1 = h = q_1{}^{r_1} \cdots q_m{}^{r_m}$ where the q_i are distinct primes. The multiplicative group of $GF(p^n)$ has order h and therefore our task is to find an element of order h in this group. Set $h_i = h/q_i$. Since $h_i < h$ there is a nonzero element $b_i \in GF(p^n)$ which is not a root of $x^{h_i} - 1$. Let $a_i = b_i^{h/q_i{}^{r_i}}$ and $a = a_1 \cdots a_m$. We have $a_i{}^{q_i{}^{r_i}} = b_i{}^h = 1$ and so the order of a_i divides $q_i{}^{r_i}$. If $a_i{}^{q_i{}^{r_i-1}} = 1$ then $b_i{}^{h/q_i} = 1$, contrary to our choice of b_i. Hence the order of a_i is exactly $q_i{}^{r_i}$. Since $a^h = 1$ the order of a divides h. Suppose that this order is not h. Then there is some prime divisor of h, say q_1, which divides h but does not divide the order of a as many as r_1 times. Then we have $a^{h/q_1} = a_1^{h/q_1} \cdots a_m^{h/q_1}$. Since $q_i{}^{r_i}$ divides h/q_1 for $i = 2, \ldots, m$, we have $a_1^{h/q_1} = 1$. This implies that $q_1{}^{r_1}$, the order of a_1, divides h/q_1, which is not true. Thus a has order h and our task is completed. ∥

7. Primitive elements

Let K be a finite algebraic extension of k. An element $a \in K$ is called a *primitive element* of K (with respect to k) if $K = k(a)$. There are instances when K does not have a primitive element (see Exercise 43 and also Exercise 44). This section will be taken up with the proof of the following result.

THEOREM 24. *Every finite separable extension of k has a primitive element.*

Proof. First, suppose that k is a finite field. Then K is also a finite field and so by Theorem 23 its multiplicative group is cyclic. Let a be a generator of the multiplicative group of K. Then $K = k(a)$.

From now on we shall assume that k has infinitely many elements. We shall show that *if there are only a finite number of fields L such that $k \subseteq L \subseteq K$ then K has a primitive element.* Let a and b be elements of K and consider the fields $k(a + bc)$ where c runs through k. For each c, $k(a + bc) \subseteq k(a,b)$. Since k has infinitely many elements our assumption concerning the subfields of K implies that there are nonzero elements c, $d \in k$ with $c \neq d$ such that $k(a + bc) = k(a + bd)$. Then $a + bd \in k(a + bc)$; hence $(a + bc) - (a + bd) = (c - d)b \in k(a + bc)$; hence $b \in k(a + bc)$. But then $a \in k(a + bc)$ and $k(a,b) = k(a + bc)$. Therefore $k(a,b)$ is a simple extension of k. Since K/k is finite, we can apply this fact a finite number of times and conclude that K is a simple extension of k.

To complete the proof of Theorem 24 we shall show that *if K is a finite separable extension of k then there are only a finite number of fields between* *k and K*. By Theorem 19 there is a smallest finite normal extension F of k such that $k \subseteq K \subseteq F$. It follows from the proof of Theorem 19 that F/k is separable. Consider the group $G(F/k)$: by the corollary to Theorem 20 this is a finite group of order $[F:k]$. Let L be a field between k and F. The group $G(F/L)$ is a subgroup of $G(F/k)$. Let M be another field between k and F with $M \neq L$. Without loss of generality we may assume that there is an element $a \in M$ with $a \notin L$. Since F/L is separable there is an element $\sigma \in G(F/L)$ such that $\sigma(a) \neq a$. This shows that $G(F/L) \neq G(F/M)$ and so that there is a one-one mapping $L \to G(F/L)$ from the set of all fields between k and F into the set of all subgroups of $G(F/k)$. Since $G(F/k)$ has only a finite number of subgroups there is only a finite number of fields between k and F. Since any field between k and K is also between k and F, there is only a finite number of fields between k and K. This completes the proof of Theorem 24. ‖

8. Algebraically closed fields

The so-called Fundamental Theorem of Algebra states that if $f(x)$ is a nonconstant polynomial in $C[x]$ where C is the field of complex numbers, $f(x)$ has a root in C. It follows immediately that every polynomial in $C[x]$ splits in $C[x]$. This is clearly equivalent to the fact that C has no proper algebraic extensions, that is, algebraic extensions which do not coincide with C. In this section we shall show that every field has an algebraic extension which shares this property with C.

DEFINITION. A field k is said to be *algebraically closed* if it has no proper algebraic extensions. A field K is called an *algebraic closure* of a field k if K/k is algebraic and if K is algebraically closed.

THEOREM 25. *The field k has an algebraic closure and any two algebraic closures of k are k-isomorphic.*

There are a number of proofs of this theorem and all known to the author require some form of the Axiom of Choice.[1] Van der Waerden presents a proof for the case when k is a countable field.[2] Of course, his proof does not require the Axiom of Choice and it can be modified to give a proof in the general case by making use of the Well-Ordering Theorem. The neatest proofs seem to be the ones based on Zorn's lemma and we prefer the one given by Zariski and Samuel.[3]

---

[1] For the various forms of the Axiom of Choice, see Chapter 0 of [47]. Here, as elsewhere in this book, the number in square brackets refers to that entry in the bibliography at the end of the book.
[2] Section 62 of [17].
[3] Section 14, Chapter II, Volume I of [18].

We shall first show that there exists an algebraic closure of k. Let $\mathfrak{N}$ be the set of all ordered pairs $(f(x),n)$ where $f(x) \in k[x]$ and n is a non-negative integer. We identify the element a of k with the element $(x - a, 0)$ of $\mathfrak{N}$. Thus the set k is a subset of $\mathfrak{N}$ and there are defined on this subset of $\mathfrak{N}$ operations with respect to which it is a field, namely, the original operations of the field k. Let $\mathfrak{S}$ be the family of all fields A such that

(i) the underlying set of A is a subset of $\mathfrak{N}$,

(ii) A is an extension of k, and

(iii) if $a = (f(x),n) \in A$ then $f(a) = 0$.

Since k satisfies (i)–(iii), $\mathfrak{S}$ is not empty. If A, $B \in \mathfrak{S}$ we write $A \leqslant B$ if A is a subfield of B. This makes $\mathfrak{S}$ into a partially ordered set and it is easy to see that every linearly ordered subset of $\mathfrak{S}$ has an upper bound in $\mathfrak{S}$: hence by Zorn's lemma, $\mathfrak{S}$ has a maximal element K.

It follows from (iii) that K/k is algebraic; we shall show that it is algebraically closed. Assuming that it is not, then there is a proper algebraic extension F of K. Define ϕ from F into $\mathfrak{N}$ in the following way. If $a \in K$ set, $\phi(a) = a$. Let $a \in F$ but $a \notin K$, then a is algebraic over k and we let $p(x) = $ Irr (k,a). Let $a_1, \ldots, a_r$ be the roots of $p(x)$ in K and $b_1, \ldots, b_s$ the roots of $p(x)$ in F but not in K. Let $n_1, \ldots, n_s$ be non-negative integers such that $a_i \neq (p(x),n_j)$ for $i = 1, \ldots, r$ and $j = 1, \ldots, s$. Set $\phi(b_j) = (p(x),n_j)$. The mapping ϕ is clearly *one-one* and maps F onto a subset F' of $\mathfrak{N}$ which contains k. Also, ϕ maps each element of K onto itself. We can make F' into a field by requiring that ϕ be an isomorphism. Then $F' \in \mathfrak{S}$ and $K \leqslant F'$ but $K \neq F'$, which contradicts the maximality of K. Thus K is algebraically closed.

Assume that K and K' are algebraic closures of k. Let $\mathfrak{M}$ be the set of all ordered triples (L,L',ϕ) where $k \subset L \subseteq K$, $k \subseteq L' \subseteq K'$, and ϕ is a k-isomorphism from L onto L'. The set $\mathfrak{M}$ is not empty since $(k,k,$ identity mapping$) \in \mathfrak{M}$. Write $(L,L',\phi) \leqslant (L_1,L_1',\phi_1)$ if $L \subseteq L_1$, $L' \subseteq L_1'$, and ϕ_1 is an extension of ϕ to L_1. This makes $\mathfrak{M}$ into a partially ordered set which, by Zorn's lemma, has a maximal element (F,F',ψ). If $F = K$ then $F' = K'$, since an isomorphic image of an algebraically closed field is clearly algebraically closed, and therefore K and K' are k-isomorphic.

Suppose $F \neq K$ and let $a \in K$ but $a \notin F$. Let $p(x) = $ Irr (F,a). Then $(\psi p)(x)$ is irreducible in $F'[x]$ and has a root $a' \in K'$. The isomorphism ψ can be extended to an isomorphism σ of $F(a)$ onto $F'(a')$. Then $(F(a), F'(a'),\sigma) \in \mathfrak{M}$ and is strictly greater than (F,F',ψ), which contradicts the maximality of (F,F',ψ) in $\mathfrak{M}$. Thus we must have $F = K$ and the proof of Theorem 25 is complete. $\parallel$

Let K be an algebraic closure of k. Then K_s, the separable closure of k in K, is called a *separable closure* of k. It follows immediately that two separable closures of k are k-isomorphic.

9. Norms and traces

Let K be a finite extension of k. By Theorem 19 there is a finite normal extension F of k such that $k \subseteq K \subseteq F$. Let n_0 be the degree of separability of K over k. By Theorem 20 there are exactly n_0 k-isomorphisms $\sigma_1, \ldots, \sigma_{n_0}$ of K onto subfields of F. If $a \in K$ we define the *norm* of a to be

$$N_{K/k}(a) = \left(\prod_{i=1}^{n_0} \sigma_i(a) \right)^{[K:k]_i}$$

and the *trace* of a to be

$$T_{K/k}(a) = [K:k]_i \sum_{i=1}^{n_0} \sigma_i(a).$$

THEOREM 26. *Let $a \in K$ and let*

$$p(x) = \mathrm{Irr}\,(k,a) = x^r + c_{r-1}x^{r-1} + \cdots + c_0.$$

Then

$$N_{K/k}(a) = ((-1)^r c_0)^{[K:k(a)]}, \qquad T_{K/k}(a) = -[K:k(a)]c_{r-1}.$$

Proof. Let r_0 be the degree of separability of $p(x)$. By Theorem 13, $r_0 = [k(a):k]_s$, and so by Theorem 20 there are r_0 k-isomorphisms $\tau_1, \ldots, \tau_{r_0}$ of $k(a)$ into F. By Theorem 18 each τ_i can be extended to a k-automorphism of F which we also denote by τ_i. If $s_0 = [K:k(a)]_s$ then by Theorem 20 there are s_0 $k(a)$-isomorphisms $\rho_1, \ldots, \rho_{s_0}$ of K into F. Then the $r_0 s_0 = n_0 = [K:k]_s$ mappings $\tau_i \rho_j$, $i = 1, \ldots, r_0$, $j = 1, \ldots, s_0$, are the n_0 k-isomorphisms of K into F. Thus

$$N_{K/k}(a) = \left(\prod_{i=1}^{r_0} \prod_{j=1}^{s_0} \tau_i \rho_j(a) \right)^{[K:k]_i}$$

$$= \left(\prod_{i=1}^{r_0} \tau_i(a) \right)^{s_0[K:k]_i}$$

and

$$T_{K/k}(a) = [K:k]_i \sum_{i=1}^{r_0} \sum_{j=1}^{s_0} \tau_i \rho_j(a)$$

$$= s_0[K:k]_i \sum_{i=1}^{r_0} \tau_i(a).$$

The polynomial $p(x)$ splits in $F[x]$ and its roots are $\tau_i(a)$, $i = 1, \ldots, r_0$. By Theorem 14,

$$p(x) = \left(\prod_{i=1}^{r_0} (x - \tau_i(a)) \right)^{[k(a):k]_i}$$

therefore,

$$c_0 = (-1)^r \left(\prod_{i=1}^{r_0} \tau_i(a) \right)^{[k(a):k]_i}$$

and

$$c_{r-1} = -[k(a):k]_i \sum_{i=1}^{r_0} \tau_i(a).$$

Therefore,

$$N_{K/k}(a) = ((-1)^r c_0)^{s_0[K:k(a)]_i} = ((-1)^r c_0)^{[K:k(a)]}$$

while

$$T_{K/k}(a) = -s_0[K:k(a)]_i c_{r-1} = -[K:k(a)]c_{r-1}. \;\|$$

COROLLARY 1. *Both $N_{K/k}(a)$ and $T_{K/k}(a)$ are elements of k.*

COROLLARY 2. *Both $N_{K/k}(a)$ and $T_{K/k}(a)$ depend only on k, K, and a, and not on the normal extension F of k used in their definitions.*

THEOREM 27. *Let K be a finite extension of k. Then*
   (1) $N_{K/k}(ab) = N_{K/k}(a)N_{K/k}(b)$ *for all a, $b \in K$,*
   (2) *if $a \in k$ then $N_{K/k}(a) = a^{[K:k]}$,*
   (3) *if σ is an isomorphism from K onto some field $\sigma(K)$ and if $a \in K$ then*
$N_{\sigma(K)/\sigma(k)}(\sigma(a)) = \sigma(N_{K/k}(a))$,
   (4) *if L is a finite extension of K and if $a \in L$ then $N_{L/k}(a) = N_{K/k}(N_{L/K}(a))$,*
   (5) $T_{K/k}(a + b) = T_{K/k}(a) + T_{K/k}(b)$ *for all a, $b \in K$,*
   (6) *if $a \in k$ then $T_{K/k}(a) = [K:k]a$,*
   (7) *if σ is an isomorphism from K onto some field $\sigma(K)$ and if $a \in K$ then*
$T_{\sigma(K)/\sigma(k)}(\sigma(a)) = \sigma(T_{K/k}(a))$,
   (8) *if L is a finite extension of K and if $a \in L$ then $T_{L/k}(a) = T_{K/k}(T_{L/K}(a))$.*

Proof. We shall prove (1)–(4). The proofs of (5)–(8) are similar and are left as an exercise. Let F be a finite normal extension of k such that $k \subseteq K \subseteq F$, let $n_0 = [K:k]_s$, and let $\sigma_1, \ldots, \sigma_{n_0}$ be the k-isomorphisms of K into F. For a, $b \in K$ we have

$$N_{K/k}(ab) = \left(\prod_{i=1}^{n_0} \sigma_i(ab) \right)^{[K:k]_i}$$

$$= \left(\prod_{i=1}^{n_0} \sigma_i(a) \right)^{[K:k]_i} \left(\prod_{i=1}^{n_0} \sigma_i(b) \right)^{[K:k]_i}$$

$$= N_{K/k}(a)N_{K/k}(b).$$

If $a \in k$ then $\sigma_i(a) = a$ for $i = 1, \ldots, n_0$ and (2) follows from the fact that $[K:k] = n_0[K:k]_i$.

To prove (3) we assume that F is the smallest normal extension of k which contains K and we let F_1 be the smallest normal extension of $\sigma(k)$ which contains $\sigma(K)$. In a natural way we can extend σ to an isomorphism from F onto F_1 which we also denote by σ. We have $n_0 = [\sigma(K):\sigma(k)]_s$ and the

n_0 $\sigma(k)$-isomorphisms of $\sigma(K)$ into F_1 are given by $\sigma\sigma_i\sigma^{-1}$, $i = 1, \ldots, n_0$.
Then for $a \in K$ we have

$$N_{\sigma(K)/\sigma(k)}(\sigma(a)) = \left(\prod_{i=1}^{n_0}\sigma\sigma_i\sigma^{-1}(\sigma(a))\right)^{[\sigma(K):\,\sigma(k)]_i}$$

$$= \left(\prod_{i=1}^{n_0}\sigma\sigma_i(a)\right)^{[K:k]_i}$$

$$= \sigma\left(\left(\prod_{i=1}^{n_0}\sigma_i(a)\right)^{[K:k]_i}\right)$$

$$= \sigma(N_{K/k}(a)).$$

Now let L be a finite extension of K and let F be a finite normal extension of k such that $k \subseteq K \subseteq L \subseteq F$. Let $n_0 = [K:k]_s$ and $m_0 = [L:K]_s$: then $n_0\,m_0 = [L:k]_s$. Let $\sigma_1, \ldots, \sigma_{n_0}$ be the k-isomorphisms from K into F and $\tau_1, \ldots, \tau_{m_0}$ the K-isomorphisms of L into F. Each σ_i can be extended to a k-automorphism of F which we also denote by σ_i. Then $\sigma_i\tau_j$, $i = 1, \ldots, n_0$, $j = 1, \ldots, m_0$, are the k-isomorphisms of L into F. Thus, if $a \in L$, we have

$$N_{L/k}(a) = \left(\prod_{i=1}^{n_0}\prod_{j=1}^{m_0}\sigma_i\tau_j(a)\right)^{[L:k]_i}$$

$$= \left(\prod_{i=1}^{n_0}\sigma_i\left(\left(\prod_{j=1}^{m_0}\tau_j(a)\right)^{[L:K]_i}\right)^{[K:k]_i}\right)$$

$$= N_{K/k}(N_{L/K}(a)). \;\|$$

Let K be a finite extension of k with $[K:k] = n$ and let $a_1, \ldots, a_n$ be a basis of K over k. The determinant of the matrix $[T_{K/k}(a_ia_j)]$ is called the *discriminant* of this basis. If $b_1, \ldots, b_n$ is another basis of K over k then there is a nonsingular $n \times n$ matrix A over k such that

$$\begin{bmatrix} b_1 \\ \cdot \\ \cdot \\ \cdot \\ b_n \end{bmatrix} = A \begin{bmatrix} a_1 \\ \cdot \\ \cdot \\ \cdot \\ a_n \end{bmatrix}.$$

Then

$$\det\,[T_{K/k}(b_ib_j)] = (\det A)^2\,\det\,[T_{K/k}(a_ia_j)]:$$

this follows from Theorem 27(5) and Exercise 50. Therefore, if one basis of K/k has nonzero discriminant the same is true of all bases of K/k. We say that K/k has *nonzero discriminant* or *zero discriminant* according as each basis of K/k has nonzero or zero discriminant.

THEOREM 28. *The extension K/k has zero discriminant if and only if $T_{K/k}(a) = 0$ for all $a \in K$.*

Proof. The sufficiency of the condition is obvious. On the other hand, suppose that K/k has zero discriminant. If $a_1, \ldots, a_n$ is a basis of K/k then $\det [T_{K/k}(a_i a_j)] = 0$. Hence the columns of $[T_{K/k}(a_i a_j)]$ are linearly dependent over k and so we can find elements $c_1, \ldots, c_n \in k$, not all zero, such that

$$\sum_{j=1}^{n} c_j T_{K/k}(a_i a_j) = 0 \qquad \text{for all } i.$$

Let $c = c_1 a_1 + \cdots + c_n a_n$. Then $c \neq 0$ and for $i = 1, \ldots, n$,

$$T_{K/k}(a_i c) = T_{K/k}\left(a_i \sum_{j=1}^{n} c_j a_j\right),$$

$$= \sum_{j=1}^{n} c_j T_{K/k}(a_i a_j) = 0,$$

where we have used the result of Exercise 50. Let $a \in K$. Since $c \neq 0$ there is an element $d \in K$ such that $a = cd$, and we write $d = d_1 a_1 + \cdots + d_n a_n$ where $d_1, \ldots, d_n \in k$. Then

$$T_{K/k}(a) = T_{K/k}\left(\sum_{i=1}^{n} d_i a_i c\right),$$

$$= \sum_{i=1}^{n} d_i T_{K/k}(a_i c) = 0. \parallel$$

THEOREM 29. *The extension K/k has nonzero discriminant if and only if K/k is separable.*

Proof. Suppose that K/k is inseparable. Then $\operatorname{char} k = p$ and $[K:k]_i = p^e$ for some positive integer e. Therefore, from the definition of trace, we have $T_{K/k}(a) = 0$ for all $a \in K$. Thus K/k has zero discriminant.

On the other hand, suppose that K/k is separable. By Theorem 24, K has a primitive element (with respect to k), say $K = k(a)$. If $[K:k] = n$ the elements $1, a, \ldots, a^{n-1}$ form a basis of K over k. Let F be a finite normal extension of k which contains K. Let $a_1 = a, a_2, \ldots, a_n$ be the conjugates of a in F. Then for any i and j with $0 \leqslant i, j \leqslant n - 1$,

$$T_{K/k}(a^i a^j) = \sum_{r=1}^{n} a_r{}^i a_r{}^j.$$

Hence the discriminant of this basis is

$$\left(\det \begin{bmatrix} 1 & 1 & \cdots & 1 \\ a_1 & a_2 & \cdots & a_n \\ a_1{}^2 & a_2{}^2 & \cdots & a_n{}^2 \\ \cdot & \cdot & \cdots & \cdot \\ a_1{}^{n-1} & a_2{}^{n-1} & \cdots & a_n{}^{n-1} \end{bmatrix} \right)^2$$

Since $a_1, \ldots, a_n$ are distinct this Vandermonde determinant is not zero. Therefore K/k has nonzero discriminant. ‖

EXERCISES

Section 1

1. Let K be an extension of k. Show that $K = k$ if and only if $[K:k] = 1$. Show also that if $[K:k]$ is a prime then there is no field L such that $k \subset L \subset K$.

2. Let D be an integral domain that contains a field k as a subring and such that the identity elements of k and D coincide. Then D is a vector space over k. Show that if D has finite dimension over k then D is a field.

Section 2

3. Let K be an algebraic extension of k and L an algebraic extension of K. Show that L is an algebraic extension of k.

4. Let K be an extension of k and a an element of K which is algebraic over k. Let $n = \deg \text{Irr } (k,a)$. Show that a has at most n conjugates in K.

5. Let $k(x)$ be the field of rational functions over the field k. Let $a = x^3/(x + 1)$. Show that $k(x)$ is an algebraic extension of $k(a)$. What is $[k(x):k(a)]$?

6. The polynomial $x^4 + 1$ is irreducible in $Q[x]$. Let $K = Q(a)$ where a is a root of this polynomial. Factor $x^4 + 1$ into irreducible factors in $K[x]$.

7. Let k be a field with two elements. Find an irreducible polynomial of degree 2 in $k[x]$ and adjoin one of its roots to k to obtain a field having four elements. By the same technique obtain a field having eight elements.

8. Prove the following extension of the last part of Theorem 5. Let k and K be fields and let σ be an isomorphism from k onto K. Let $p(x)$ be a nonconstant irreducible polynomial in $k[x]$. Then $(\sigma p)(x)$ is irreducible in $K[x]$ (prove this). Let a be a root of $p(x)$ and b a root of $(\sigma p)(x)$. Then there is a unique isomorphism τ from $k(a)$ onto $K(b)$ such that $\tau(a) = b$ and $\tau(c) = \sigma(c)$ for all $c \in k$.

9. Let K be a finite extension of k, let $a \in K$, and let $n = \deg \text{Irr } (k,a)$. Show that n divides $[K:k]$.

10. Let $K = Q(\sqrt{2},\sqrt{3})$. Determine $[K:Q]$ and find a basis of K over Q.

11. Do the same for $K = Q(i,\sqrt{3},a)$ where a is a complex cube root of unity.

12. Let $p(x) = x^3 - x + 1$ and let a be a root of $p(x)$. Find the inverse of $1 - 2a + 3a^2$ in $Q(a)$.

13. Let $p(x)$ and a be as in Exercise 12. Let $b = 1 + a - 2a^2$. Find Irr (Q,b).

14. Let R be the field of real numbers. Exhibit an isomorphism from $R[x]/(x^2 + 1)$ onto the field of complex numbers.

15. Let K be a simple transcendental extension of k. Show that K is k-isomorphic to $k(x)$, the field of rational functions over k.

Section 3

16. Let k be a field of characteristic p. Show that for all $a, b \in k$ and all positive integers n we have $(a + b)^{p^n} = a^{p^n} + b^{p^n}$.

17. Let $f(x) \in k[x]$. Show that a is a root of multiplicity m of $f(x)$ if and only if $f^{(j)}(a) = 0$ for $j = 0, 1, \ldots, m - 1$, but $f^{(m)}(a) \neq 0$.

18. Let K be a separable extension of k and let $k \subseteq L \subseteq K$. Show that K is a separable extension of L.

19. Show that k is perfect if and only if it has no finite inseparable extensions.

20. Let k be a field of characteristic p and let $k(x)$ be the field of rational functions over k. Show that $k(x)$ is not perfect by showing that there is no element $z \in k(x)$ such that $z^p = x$.

21. Let K be an algebraic extension of a perfect field. Show that K is perfect.

22. Let K be an extension of k and assume that K is perfect. Describe the smallest perfect extension of k in K.

Section 4

23. Let K be an extension of k and let $a \in K$ be purely inseparable over k. Show that $k(a)$ is a purely inseparable extension of k.

24. Let K be a finite extension of a field k and assume that k is not perfect. Show that K is not perfect.

25. Let x and y be indeterminates over a field k of characteristic p and let $K = k(x^{1/p}, y^{1/p})$. Show that $[K : k(x,y)] = p^2$ and that $a^p \in k(x,y)$ for all $a \in K$.

26. Let k be a field of characteristic p. A subset B of k is called a *p-base* of k if $k = k^p(B)$ and for every finite subset $\{b_1, \ldots, b_r\}$ of B we have $[k^p(b_1, \ldots, b_r) : k^p] = p^r$. Show that

(a) k has a p-base,
(b) if B is a p-base of k and if n is a positive integer then $k = k^{p^n}(B)$, and
(c) if B is a p-base of k then for every finite subset $\{b_1, \ldots, b_r\}$ of B and for every positive integer n we have $[k^{p^n}(b_1, \ldots, b_r) : k^{p^n}] = p^{nr}$.

Section 5

27. If $k \subseteq L \subseteq K$ and if K is a normal extension of k, show that K is a normal extension of L.

28. Describe the splitting field of $x^3 - 2$ over the rational field Q. Show that if a is any root of this polynomial then $Q(a)/Q$ is not normal.

29. Show that every quadratic extension of k is normal over k.

30. Let char $k = p$ and let b be a root of $x^p - x - a \in k[x]$. Show that $k(b)/k$ is normal. Is it separable? Find necessary and sufficient conditions for this polynomial to be irreducible in $k[x]$.

31. If $k \subseteq L \subseteq K$ with K/k finite, show that $[K:k]_s = [K:L]_s[L:k]_s$ and that $[K:k]_i = [K:L]_i[L:k]_i$.

32. Let $k \subseteq L \subseteq F \subseteq K$ with K/k finite normal. If there are exactly n L-isomorphisms from F into K, show that each k-isomorphism from L into K has exactly n extensions to k-isomorphisms from F into K.

33. Let $k \subseteq L \subseteq K$ with K/k finite normal. Let σ be a k-isomorphism from L into K. Show that σ has exactly $[K:k]_s/[L:k]_s$ extensions to k-automorphisms of K.

34. Show that a finite extension K/k is normal if and only if there is a normal extension F of k such that $k \subseteq K \subseteq F$ and every k-isomorphism from K into F is a k-automorphism of K.

35. Let a be a rational number which is not the cube of a rational number. Let K be a splitting field of $x^3 - a$ over Q. What is $[K:Q]$?

36. Let char $k \neq 2$. Show that the splitting field over k of $x^4 - (a + b)x^2 + ab$, where $a, b \in k$, has degree 4 over k if and only if $a, b,$ and ab are nonsquares in k.

37. Let $f(x) \in k[x]$ and let K be a splitting field of $f(x)$ over k. Show that if F is an extension of k then FK is a splitting field of $f(x)$ over F.

38. Using Theorems 18 and 19 and the remark between them, show that if K/k is finite then $G(K/k)$ is finite.

Section 6

39. Show that in $GF(p^n)$ each element has exactly one pth root.

40. Construct a field having 25 elements.

41. Let K be a splitting field over $GF(p)$ of the polynomial $x^{p^n} - x \in GF(p)[x]$. Show that every element of K is a root of this polynomial and that K has p^n elements. This shows the existence of a field with p^n elements.

42. Show that $GF(p^m)$ is (isomorphic to) a subfield of $GF(p^n)$ if and only if m divides n. Show that in this case $GF(p^n)$ has exactly one subfield with p^m elements.

Section 7

43. Let x and y be indeterminates over a field of characteristic p. Show that $K = k(x^{1/p}, y^{1/p})$ has no primitive element (with respect to $k(x,y)$). In connection with this exercise and with Exercise 25, see Exercise 44.

44. Let k be a field of characteristic p and let K be a finite extension of k. Let e be the smallest non-negative integer such that $a^{p^e} \in k$ for all $a \in K$ (assuming such an integer exists). Let $[K:k] = p^f$. Show that K has a primitive element (with respect to k) if and only if $e = f$.

45. Let a and b be nonsquare integers. Describe $Q(\sqrt{a}) \cap Q(\sqrt{b})$. Find a primitive element of the composite $Q(\sqrt{a})Q(\sqrt{b})$ with respect to Q.

Section 8

46. Prove that no finite field is algebraically closed.

47. Construct a proof of the existence of an algebraic closure of a field k by well-ordering $k[x]$ and adjoining the roots of these polynomials in order.

Section 9

48. Let K/k be finite and let $e_1, \ldots, e_n$ be a basis of K/k. For $a \in K$ let

$$ae_i = \sum_{j=1}^{n} a_{ij}e_j, \qquad a_{ij} \in k, \qquad i = 1, \ldots, n.$$

Let $A = [a_{ij}]$. Show that $N_{K/k}(a) = \det A$, and $T_{K/k}(a) =$ the trace of $A = a_{11} + \cdots + a_{nn}$.

49. Carry out the proofs of parts (5)–(8) of Theorem 27.

50. Let K/k be finite and let $a \in k$ and $b \in K$. Show that $T_{K/k}(ab) = aT_{K/k}(b)$.

51. Let char $k = p$ and let K/k be of finite degree n. Show that if K/k has zero discriminant then p divides n.

52. Let $K = k(a)$ be a separable extension of k and let $f(x) = \mathrm{Irr}\,(k,a)$ where $\deg f(x) = n$. Show that the discriminant of the basis $1, a, \ldots, a^{n-1}$ of K/k is given by

$$(-1)^{n(n-1)/2}N_{K/k}(f'(a)).$$

Additional exercises

53. Let K and L be extensions of k, both contained in some extension C of k. We say that K is *linearly disjoint* from L over k if every finite set of elements of K which is linearly independent over k is still linearly independent over L. Show that this is a symmetric relation on the set of all extensions of k in C.

54. Show that K and L are linearly disjoint over k if and only if the map $a \otimes b \to ab$ from $K \otimes_k L$ into the vector space over k generated by $\{ab \mid a \in K, b \in L\}$ is a k-isomorphism.

55. With k, K, and L as above, let $k \subseteq E \subseteq L$. Show that K and L are linearly disjoint over k if and only if K and E are linearly disjoint over k and KE and L are linearly disjoint over E.

56. We say that K is *free* from L over k if every finite set of elements of K which is algebraically independent over k remains so over L. Show that this is a symmetric relation on the set of extensions of k which are contained in C.

57. Show that if K and L are linearly disjoint over k, they are free over k. Give an example of extensions K and L of k which are free but not linearly disjoint over k.

58. Let $k \subseteq L \subseteq K$ and let $a_1, \ldots, a_n$ be elements of K which are algebraically independent over k. Show that L and $k(a_1, \ldots, a_n)$ are linearly disjoint over k.

59. Let K be an algebraic extension of k and let C be an algebraic closure of k which contains K. Let L be the set of elements of C left fixed by each element of $G(C/k)$. Show that K/k is separable if and only if K and L are linearly disjoint over k.

60. Let char $k = p$ and let C be an algebraic closure of k. Let $k^{p^{-\infty}}$ be the smallest perfect field in C which contains k (see Exercise 22) and let $k^{p^{-1}} = \{a \in C \mid a^p \in k\}$. Let K be an extension of k in C. Show that if K/k is separable then K and $k^{p^{-\infty}}$ are linearly disjoint over k. Show that if K and $k^{p^{-1}}$ are linearly disjoint over k then K/k is separable.

CHAPTER 2

Galois Theory

1. Automorphisms of extensions: Galois extensions

Let k be a field and let K be a finite extension of k. In this chapter we shall make a detailed study of the group of k-automorphisms of K and of the k-isomorphisms of K into normal extensions of k. The first part of our discussion will culminate in the Fundamental Theorem of Galois Theory. We shall then give some examples and study some special types of extensions. Finally, in the last three sections of this chapter, we shall drop the assumption that K is a finite extension of k and study the group of k-automorphisms of a possibly infinite extension of k.

THEOREM 1. *Let K and L be fields and let $\sigma_1, \ldots, \sigma_n$ be distinct isomorphisms of K into L. Then $\sigma_1, \ldots, \sigma_n$ are linearly independent in the sense that if*

$$\sum_{i=1}^{n} a_i \sigma_i(b) = 0, \qquad a_1, \ldots, a_n \in L,$$

for all $b \in K$ then $a_1 = \cdots = a_n = 0$.

Proof. By induction on n. Suppose that the relation in the statement of the theorem holds for all $b \in K$. If $n = 1$ we have $a_1\sigma_1(b) = 0$ for all $b \in K$ and if we take $b = 1$ we get $a_1 = 0$. Now suppose that $n > 1$ and that the result of the theorem holds for any $n - 1$ isomorphisms of K into L. If some $a_i = 0$, say $a_1 = 0$, then

$$\sum_{i=2}^{n} a_i \sigma_i(b) = 0$$

for all $b \in K$ and we have $a_2 = \cdots = a_n = 0$ by our induction assumption. Hence either the theorem is true or $a_i \neq 0$ for $i = 1, \ldots, n$. Suppose the latter. Since σ_1 and σ_n are distinct there is an element $c \in K$ such that $\sigma_1(c) \neq \sigma_n(c)$. Then, since $a_n \neq 0$, we have

$$a_n^{-1}\sigma_n(c^{-1}) \sum_{i=1}^{n} a_i\sigma_i(cb) = 0$$

for all $b \in K$, or

$$\sum_{i=1}^{n} a_i a_n^{-1}\sigma_n(c^{-1})\sigma_i(c)\sigma_i(b) = 0$$

for all $b \in K$. If we subtract this from

$$a_n^{-1} \sum_{i=1}^{n} a_i\sigma_i(b) = 0$$

we obtain

$$\sum_{i=1}^{n-1} a_i a_n^{-1}(1 - \sigma_n(c^{-1})\sigma_i(c))\sigma_i(b) = 0,$$

which must hold for all $b \in K$. By our induction assumption this implies that $a_i a_n^{-1}(1 - \sigma_n(c^{-1})\sigma_i(c)) = 0$ for $i = 1, \ldots, n-1$. In particular, $1 - \sigma_n(c^{-1})\sigma_1(c) = 0$ or $\sigma_1(c) = \sigma_n(c)$, which is contrary to our choice of c. ‖

COROLLARY. *Let K be an extension of k and let $\sigma_1, \ldots, \sigma_n$ be distinct k-automorphisms of K. Then $\sigma_1, \ldots, \sigma_n$ are linearly independent in the sense of the theorem.*

THEOREM 2. *Let K and L be fields and let $\sigma_1, \ldots, \sigma_n$ be distinct isomorphism from K into L. Let*

$$F = \{a \in K \mid \sigma_1(a) = \cdots = \sigma_n(a)\}.$$

Then F is a subfield of K and $[K:F] \geqslant n$.

Proof. Clearly F is a subfield of K. Assume that $[K:F] = r < n$ and let $b_1, \ldots, b_r$ be a basis of K over F. Consider the system of r equations in n unknowns

$$\sum_{j=1}^{n} \sigma_j(b_i)x_j = 0, \qquad i = 1, \ldots, r.$$

Here we have more unknowns than equations and therefore the system of equations has a nontrivial solution $c_1, \ldots, c_n$ in L. Let a be an arbitrary element of K and write $a = a_1b_1 + \cdots + a_rb_r$ where $a_1, \ldots, a_r \in F$. Multiply the ith one of the preceding equations by $\sigma_1(a_i) = \cdots = \sigma_n(a_i)$ and obtain

$$\sum_{j=1}^{n} \sigma_j(a_ib_i)c_j = 0.$$

If we add these equations we obtain

$$\sum_{j=1}^{n} \left(\sum_{i=1}^{r} \sigma_j(a_ib_i) \right) c_j = 0,$$

or

$$\sum_{j=1}^{n} c_j \sigma_j(a) = 0.$$

Since this holds for all $a \in K$ and not all of the c_j are zero, we have contradicted Theorem 1. ‖

Let K be a finite extension of k. As in Chapter 1 we let $G(K/k)$ denote the group of k-automorphisms of K. Let H be a subgroup of $G(K/k)$ and let

$$F(H) = \{a \in K \mid \sigma(a) = a \text{ for all } \sigma \in H\}.$$

By Theorem 2, $F(H)$ is a subfield of K, which clearly contains k, and $[K:F(H)] \geqslant \#H$, where $\#H$ denotes the order of H. The field $F(H)$ is called the *fixed field* of H.

As an example of how we can use what has been proved, consider the field $K = k(x)$ where x is an indeterminate, that is, K is the field of rational functions over k. Each of the mappings of K into itself given by

$$f(x) \rightarrow f(x), \qquad \text{the identity mapping,}$$

$$f(x) \rightarrow f(1 - x),$$

$$f(x) \rightarrow f\left(\frac{1}{x}\right),$$

$$f(x) \rightarrow f\left(1 - \frac{1}{x}\right),$$

$$f(x) \rightarrow f\left(\frac{1}{1 - x}\right),$$

$$f(x) \rightarrow f\left(\frac{x}{x - 1}\right),$$

is an automorphism of K. Let F be the set of elements of K which are left fixed by each of these automorphisms. By Theorem 2, F is a field and $[K:F] \geqslant 6$.

We shall now show that $[K:F] = 6$. Let

$$g(x) = \frac{(x^2 - x + 1)^3}{x^2(x - 1)^2}.$$

By performing the necessary substitutions it is a routine matter to show that $g(x) \in F$. Let $L = k(g(x))$, that is, L is the field of rational functions in $g(x)$ with coefficients in k. Then $L \subseteq F$ and $[K:L] \geqslant 6$. Now we have

$$(x^2 - x + 1)^3 - g(x)x^2(x - 1)^2 = 0,$$

so that x is a root of a polynomial of degree 6 with coefficients in L. Hence $[L(x):L] \leqslant 6$. But $L(x) = k(x) = K$ and as a result $[K:L] \leqslant 6$. Thus $[K:L] = 6$ and we conclude that $[K:F] = 6$ and $F = k(g(x))$.

THEOREM 3. *Let G be a finite group of automorphisms of a field K and let F be the set of all elements of K which are left fixed by each element of G. Then F is a field and $[K:F] = \#G$.*

Proof. By Theorem 2, F is a subfield of K and $[K:F] \geqslant \#G = n$. Assume that $[K:F] > n$. Then there are $n + 1$ elements $b_1, \ldots, b_{n+1}$ of K which are linearly independent over F. Let $G = \{\sigma_1, \ldots, \sigma_n\}$ and consider the system of n equations in $n + 1$ unknowns,

$$\sum_{j=1}^{n+1} \sigma_i(b_j)x_j = 0, \qquad i = 1, \ldots, n. \tag{1}$$

This system has more unknowns than equations and so it has a nontrivial solution $a_1, \ldots, a_{n+1}$ in K. At least one of the a_i is not in F. For, suppose each $a_i \in F$. If σ_1 is the identity element of the group G then the first of the above equations is

$$\sum_{j=1}^{n+1} a_j b_j = 0$$

which contradicts the fact that $b_1, \ldots, b_{n+1}$ are linearly independent over F.

From among all of the solutions of the system of equations (1) choose one with the smallest number of nonzero terms. Let this be $a_1, \ldots, a_{n+1}$ where we may assume that $a_j \neq 0$ for $j = 1, \ldots, r$ and $a_j = 0$ for $j = r + 1, \ldots, n + 1$ (if $r \neq n + 1$). Then $r \neq 1$, for if $r = 1$ we have, with σ_1 as identity of G, $a_1 b_1 = 0$ which implies that $a_1 = 0$. We have

$$\sum_{j=1}^{r} a_j \sigma_i(b_j) = 0, \qquad i = 1, \ldots, n. \tag{2}$$

We may assume that $a_r = 1$, for otherwise we could multiply each of the equations by a_r^{-1}. Also, we may assume that $a_1 \notin F$ so that there is an element $\sigma_h \in G$ such that $\sigma_h(a_1) \neq a_1$. If we apply σ_h to each equation in (2) we obtain

$$\sum_{j=1}^{r} \sigma_h(a_j)\sigma_h\sigma_i(b_j) = 0, \qquad i = 1, \ldots, n.$$

As σ_i runs through the group G, so does $\sigma_h\sigma_i$ and we may rewrite these equations as

$$\sum_{j=1}^{r} \sigma_h(a_j)\sigma_i(b_j) = 0, \qquad i = 1, \ldots, n. \tag{3}$$

If we subtract the ith equation in (3) from the ith equation in (2) and use the fact that $a_r = 1$ we obtain

$$\sum_{j=1}^{r-1}(a_j - \sigma_h(a_j))\sigma_i(b_j) = 0, \qquad i = 1, \ldots, n.$$

Since $a_1 - \sigma_h(a_1) \neq 0$, this contradicts our choice of the solution $a_1, \ldots, a_{n+1}$ of the system (1) as the one with the smallest number of nonzero terms. ‖

Let K be a finite normal extension of k and let F be the fixed field of $G(K/k)$. It has been shown in Chapter 1 that F/k is purely inseparable. We note that $G(K/F) = G(K/k)$ and so by Theorem 3, $[K:F] = \#G(K/F)$. However, by the corollary to Theorem 20 in Chapter 1, $[K:F]_s = \#G(K/F)$. Therefore, $[K:F]_s = [K:F]$, which implies that K/F is separable. This completes the proof of the corollary to Theorem 21 in Chapter 1. Note that $[K:F] = \#G(K/k) = [K:k]_s$ and so $[F:k] = [K:k]_i$.

DEFINITION. An algebraic extension K of k is called a *Galois extension* of k if the fixed field of $G(K/k)$ is k. In this case, $G(K/k)$ is called the *Galois group* of K/k.

THEOREM 4. *A finite extension K of k is a Galois extension of k if and only if K is a normal and separable extension of k.*

Proof. Let K be a finite Galois extension of k. Let $G(K/k) = \{\sigma_1, \ldots, \sigma_n\}$: we know from Exercise 38, Chapter 1, that this group is finite. Let $a \in K$, $a \notin k$, and let $a_1 = a, a_2, \ldots, a_r$ be the distinct elements among $\sigma_1(a), \ldots, \sigma_n(a)$. Since $G(K/k)$ is a group, if we apply one of its elements, say σ_i, to $a_j = \sigma_j(a)$ we obtain $\sigma_i(a_j) = \sigma_i\sigma_j(a) = \sigma_h(a)$ where $\sigma_h = \sigma_i\sigma_j$. Thus the elements of $G(K/k)$ permute the elements $a_1, \ldots, a_r$ of K among themselves. Therefore every symmetric function of $a_1, \ldots, a_r$ and, in particular, every coefficient of $f(x) = (x - a_1) \cdots (x - a_r)$, lies in the fixed field of $G(K/k)$, that is, in k. Let $g(x)$ be an irreducible factor of $f(x)$ in $k[x]$ such that, say, $g(a_i) = 0$. If $a_i = \sigma_i(a)$ and $a_j = \sigma_j(a)$, then $a_j = \sigma_j\sigma_i^{-1}(a_i)$ so that $g(a_j) = 0$. Thus every root of $f(x)$ is a root of $g(x)$. This implies that $f(x)$ is irreducible in $k[x]$. We have shown that every element $a \in K$ is a simple root of Irr (k,a) and that Irr (k,a) splits in $K[x]$. Thus K/k is both normal and separable.

Conversely, assume that K/k is finite, normal, and separable. Let F be the fixed field of $G(K/k)$. Then $[F:k] = [K:k]_i = 1$ and $F = k$. ‖

2. The fundamental theorem of Galois theory

The theorem referred to in the title of this section is the following one.

THEOREM 5. *Let K be a finite Galois extension of k. Then there is a one-one correspondence between the fields L such that $k \subseteq L \subseteq K$ and the subgroups* *of* $G(K/k)$. *This correspondence is given by*

$$L \leftrightarrow G(K/L).$$

If $k \subseteq L \subseteq K$ *then* L/k *is Galois if and only if* $G(K/L)$ *is a normal subgroup of* $G(K/k)$. *In this case*

$$G(L/k) \cong G(K/k)/G(K/L).$$

Proof. Define a mapping from the set of all fields between k and K into the set of all subgroups of $G(K/k)$ as follows: if $k \subseteq L \subseteq K$ we map L onto $G(K/L)$. Since K/k is Galois it is separable and we have shown in the proof of Theorem 24, Chapter 1, that in this circumstance this mapping is *one-one*. It remains to show that it is *onto*. Let H be a subgroup of $G(K/k)$ and let L be the fixed field of H. Since K/k is normal and separable, K/L is normal and separable. Hence K/L is Galois and L is the fixed field of $G(K/L)$. Each element of H leaves each element of L fixed and therefore $H \subseteq G(K/L)$. By Theorem 3, $\#H = [K:L]$. By the remarks following Theorem 3, $\#G(K/L) = [K:L]_s = [K:L]$ since K/L is separable. Hence $H = G(K/L)$ so that H is the image of L under the mapping we are considering.

Now consider the field L between k and K. Suppose that L/k is normal and let $\sigma \in G(K/k)$ and $\tau \in G(K/L)$. If $a \in L$ then each conjugate of a is in L. Thus, since $\sigma(a)$ is one of the conjugates of a, we have $\sigma^{-1}\tau\sigma(a) = \sigma^{-1}\sigma(a) = a$, that is, $\sigma^{-1}\tau\sigma \in G(K/L)$. Thus $G(K/L)$ is a normal subgroup of $G(K/k)$. On the other hand, suppose that $G(K/L)$ is a normal subgroup of $G(K/k)$. We shall show that L/k is normal (we already know that L/k is separable). Let $f(x)$ be a nonconstant irreducible polynomial in $k[x]$ which has one root, say a, in L. Since K/k is normal $f(x)$ splits in $K[x]$ and all of the roots of $f(x)$ can be expressed in the form $\sigma(a)$ for some $\sigma \in G(K/k)$. Let $\tau \in G(K/L)$. Then there is an element $\rho \in G(K/L)$ such that $\tau = \sigma\rho\sigma^{-1}$, where σ is some element of $G(K/k)$. Then $\tau\sigma(a) = \sigma\rho\sigma^{-1}\sigma(a) = \sigma\rho(a) = \sigma(a)$ and so $\sigma(a)$ is left fixed by each element of $G(K/L)$. This means that $\sigma(a) \in L$ for all $\sigma \in G(K/k)$ and that $f(x)$ splits in $L[x]$. Therefore L/k is normal.

Finally, consider the mapping ϕ from $G(K/k)$ into $G(L/k)$ given by $\phi(\sigma) =$ the restriction of σ to L. The mapping ϕ is clearly a homomorphism and is onto $G(L/k)$ by Theorem 18 of Chapter 1. The kernel of ϕ consists of all $\sigma \in G(K/k)$ such that $\sigma(a) = a$ for all $a \in L$, that is, the kernel of ϕ is precisely $G(K/L)$. Therefore, $G(L/k) \cong G(K/k)/G(K/L)$. ∥

Remark. Note that if $k \subseteq L \subseteq K$ and if $k \subseteq F \subseteq K$ then $L \subseteq F$ if and only if $G(K/L) \supseteq G(K/F)$.

THEOREM 6. *Let K and F be extensions of k with K/k finite and assume that both K and F are subfields of some larger field. If K is a Galois extension of k then KF is a Galois extension of F and*

$$G(KF/F) \cong G(K/K \cap F).$$

Proof. The field K is a splitting field over k of some polynomial $f(x) \in k[x]$ whose irreducible factors have only simple roots. Then KF is a splitting field of $f(x)$ over F. Hence KF is normal and separable, and so Galois, over F.

Let $\sigma \in G(KF/F)$. If the distinct roots of $f(x)$ are $a_1, \ldots, a_n$ then σ induces a permutation of these roots which in turn induces an element $\sigma' \in G(K/k)$. Clearly, distinct elements of $G(KF/F)$ induce in this manner distinct elements of $G(K/k)$ and the mapping $\sigma \to \sigma'$ is an isomorphism from $G(KF/F)$ onto some subgroup of $G(K/k)$. If $\sigma \in G(KF/F)$ then σ' leaves fixed each element of F that lies in K, that is, $\sigma' \in G(K/K \cap F)$. On the other hand, if $\sigma' \in G(K/K \cap F)$ then σ' permutes $a_1, \ldots, a_n$ and determines an element $\tau \in G(KF/F)$. But τ and σ' induce the same permutation of $a_1, \ldots, a_n$ and therefore $\tau \to \sigma'$. Thus the mapping $\sigma \to \sigma'$ is an isomorphism from (GKF/F) onto $G(K/K \cap F)$. ‖

DEFINITION. Let K be a Galois extension of k. We say that K is an *Abelian*, or *cyclic*, or *solvable*, etc., *extension* of k according as $G(K/k)$ is an Abelian, or cyclic, or solvable, etc., group.

3. An example

Let K be the splitting field of $f(x) = x^3 - 2$ over the field Q of rational numbers. We may consider K as a subfield of the field of complex numbers. To obtain K we first adjoin $\sqrt[3]{2}$ (real) to Q. In $Q(\sqrt[3]{2})[x]$ we have

$$f(x) = (x - \sqrt[3]{2})[x^2 + \sqrt[3]{2}\, x + (\sqrt[3]{2})^2]$$

$$= (x - \sqrt[3]{2})g(x).$$

The field $Q(\sqrt[3]{2})$ is a real field and $g(x)$ has no real roots, so that $g(x)$ is irreducible in $Q(\sqrt[3]{2})[x]$. The roots of $g(x)$ are $\sqrt[3]{2}\, \alpha$ and $\sqrt[3]{2}\, \beta$ where $\alpha = (-1 + \sqrt{3}\, i)/2$ and $\beta = (-1 - \sqrt{3}\, i)/2$. Hence

$$K = Q(\sqrt[3]{2}, \sqrt{3}\, i).$$

We have $[K:Q] = 6$ and since K/Q is a Galois extension, the order of $G(K/Q)$ is 6. Thus, if we can find six distinct Q-automorphisms of K they will constitute all of $G(K/Q)$; one element of this group is its identity element ε.

An element of $G(K/Q)$ is completely determined by what it does to $\sqrt[3]{2}$ and $\sqrt{3}\, i$. Let σ be the Q-automorphism of K which leaves $\sqrt[3]{2}$ fixed and maps $\sqrt{3}\, i$ into $-\sqrt{3}\, i$. Let τ be the Q-automorphism of K which maps $\sqrt[3]{2}$ onto $\sqrt[3]{2}\, \alpha$ and $\sqrt{3}\, i$ onto $- \sqrt{3}\, i$. We have the following table which gives the images of $\sqrt[3]{2}$ and $\sqrt{3}\,i$ under the indicated Q-automorphisms of K.

	ε	σ	τ	$\sigma\tau$	$\tau\sigma$	$\sigma\tau\sigma$
$\sqrt[3]{2}$	$\sqrt[3]{2}$	$\sqrt[3]{2}$	$\sqrt[3]{2}\,\alpha$	$\sqrt[3]{2}\,\beta$	$\sqrt[3]{2}\,\alpha$	$\sqrt[3]{2}\,\beta$
$\sqrt{3}\,i$	$\sqrt{3}\,i$	$-\sqrt{3}\,i$	$-\sqrt{3}\,i$	$\sqrt{3}\,i$	$\sqrt{3}\,i$	$-\sqrt{3}\,i$

Thus,
$$G(K/Q) = \{\varepsilon,\sigma,\tau,\sigma\tau,\tau\sigma,\sigma\tau\sigma\}.$$

The subgroups of $G(K/Q)$ are

$$H_1 = \{\varepsilon,\sigma\}, \qquad H_2 = \{\varepsilon,\tau\}, \qquad H_3 = \{\varepsilon,\sigma\tau\sigma\}, \qquad H_4 = \{\varepsilon,\sigma\tau,\tau\sigma\}.$$

The fixed fields of these subgroups are, respectively,

$$L_1 = Q(\sqrt[3]{2}), \qquad L_2 = Q(\sqrt[3]{2}\,\beta), \qquad L_3 = Q(\sqrt[3]{2}\,\alpha), \qquad L_4 = Q(\sqrt{3}\,i).$$

The extension L_4/Q is Galois, but L_i/Q is not normal for $i = 1, 2, 3$.

4. Cyclotomic fields

Let n be a positive integer. By a *primitive nth root of unity* we mean a generator of the cyclic group of all complex nth roots of unity. The complex number
$$\zeta = \exp(2\pi i/n)$$
is one primitive nth root of unity. Since ζ generates a cyclic group of order n, the other primitive nth roots of unity are the complex numbers ζ^h where $1 \leqslant h \leqslant n$ and $(h,n) = 1$. Therefore, the number of primitive nth roots of unity is given by the Euler function of n, $\phi(n)$, where

$$\phi(n) = \text{the number of integers } h \text{ such that}$$
$$1 \leqslant h \leqslant n \quad \text{and} \quad (h,n) = 1.$$

This function has, among others, the following properties:[1]

(i) $\phi(mn) = \phi(m)\phi(n)$ if $(m,n) = 1$.

(ii) $\sum_{d|n} \phi(d) = n$ (see Proposition 1, below).

(iii) $\phi(n) = \sum_{d|n} d\mu(n/d)$ where μ is the function of Möbius defined by

$$\mu(n) = \begin{cases} 1 & \text{if } n = 1, \\ (-1)^t & \text{if } n = p_1 \cdots p_t \text{ where the } p_i \text{ are distinct primes,} \\ 0 & \text{otherwise.} \end{cases}$$

(iv) $\phi(n) = n \prod_{p|n} (1 - p^{-1})$.

[1] For these properties of Euler's function, see pp. 18–21 of [48].

DEFINITION. By the *nth cyclotomic polynomial* $F_n(x)$ we mean the monic polynomial whose roots are the $\phi(n)$ complex primitive nth roots of unity, that is,

$$F_n(x) = \prod_{\substack{1 \leqslant h \leqslant n \\ (h,n)=1}} (x - \zeta^h), \qquad \zeta = \exp(2\pi i/n)$$

We shall show that $F_n(x) \in Q[x]$ and that $F_n(x)$ is irreducible.

DEFINITION. The splitting field of $F_n(x)$ over Q, which is contained in the field of complex numbers, is called the *nth cyclotomic field*. We denote it by K_n.

PROPOSITION 1. *Let d range over all of the (positive) divisors of n and for each d let q range over the $\phi(n/d)$ positive integers $\leqslant n/d$ such that $(q,n/d) = 1$. Then the integers qd are precisely the integers $1, \ldots, n$, without repetition.*

Proof. First we note that for each d and each of the corresponding q's we have $1 \leqslant qd \leqslant n$. Next we show that no two of these numbers can be equal when the d's or q's differ. For $i = 1, 2$, let $d_i \,|\, n$, $1 \leqslant q_i \leqslant n/d_i$, $(q_i, n/d_i) = 1$, and suppose that $q_1 d_1 = q_2 d_2$. Then $q_1 n/d_2 = q_2 n/d_1$ and so $q_1 \,|\, q_2$ and $q_2 \,|\, q_1$, that is, $q_1 = q_2$. But then $d_1 = d_2$. Finally, we show that every integer m with $1 \leqslant m \leqslant n$ can be written in the form qd. Let $(m,n) = d$. Then $(m/d, n/d) = 1$ and $m = qd$ with $q = m/d$. ∥

PROPOSITION 2. *For any positive integer n,*

$$x^n - 1 = \prod_{d \mid n} F_d(x).$$

Proof. Let $\zeta = \exp(2\pi i/n)$. Then the roots of $x^n - 1$ are ζ^h where $h = 1, \ldots, n$. Hence

$$x^n - 1 = \prod_{h=1}^{n} (x - \zeta^h) = \prod_{d \mid n} \prod_{\substack{1 \leqslant q \leqslant n/d \\ (q, n/d)=1}} (x - \zeta^{qd}).$$

But $\zeta^d = \exp(2\pi i/(n/d))$ is a primitive (n/d)th root of unity and therefore

$$x^n - 1 = \prod_{d \mid n} F_{n/d}(x) = \prod_{d \mid n} F_d(x). \; ∥$$

PROPOSITION 3. *For any positive integer n, $F_n(x)$ has integer coefficients.*

Proof. We shall show by induction on n that $F_n(x)$ has integer coefficients. Since $F_1(x) = x - 1$ the result is true for $n = 1$. Assume that $n > 1$ and that $F_m(x)$ has integer coefficients for $1 \leqslant m \leqslant n - 1$. Then

$$F_n(x) = \frac{x^n - 1}{\prod_{\substack{d \mid n \\ d \neq n}} F_d(x)}$$

has integer coefficients. ∥

Remark. If $f(x)$ and $g(x)$ are monic polynomials with integer coefficients, then upon dividing $f(x)$ by $g(x)$ we obtain a quotient and a remainder having integer coefficients. We have used this fact in the proof of Proposition 3, and we will use it again in the proof of Proposition 4.

PROPOSITION 4. *For any positive integer n, $F_n(x)$ is irreducible in $Q[x]$.*

Proof. Let $f(x)$ be a nonconstant monic irreducible factor of $F_n(x)$ in $Q[x]$. By Exercise 19, $f(x)$ has integer coefficients. Let ζ be a primitive nth root of unity which is a root of $f(x)$. For each positive integer m we can write

$$f(x^m) = q_m(x)f(x) + r_m(x)$$

where

$$r_m(x) = 0 \text{ or } \deg r_m(x) < \deg f(x),$$

and $r_m(x)$ has integer coefficients. Then

$$f(\zeta^m) = r_m(\zeta),$$

and since $\zeta^{n+m} = \zeta^m$,

$$r_{n+m}(\zeta) = f(\zeta^{n+m}) = f(\zeta^m) = r_m(\zeta).$$

It follows then from Theorem 3 in Chapter 1, that

$$r_{n+m}(x) = r_m(x) \qquad \text{for all } m. \tag{1}$$

If p is a prime we have
$$f(x^p) \equiv (f(x))^p (\bmod p),$$

where the notation indicates that each coefficient of the polynomial on the right is congruent modulo p to the corresponding coefficient of the polynomial on the left. Hence we have

$$f(x^p) = q_p(x)f(x) + r_p(x) \equiv (f(x))^p (\bmod p),$$
or
$$r_p(x) \equiv \{-q_p(x) + (f(x))^{p-1}\}f(x)(\bmod p).$$

But $\deg r_p(x) < \deg f(x)$ or $r_p(x) = 0$ and $f(x)$ is monic so we must have $r_p(x) \equiv 0 (\bmod p)$, that is, each coefficient of $r_p(x)$ is divisible by p.

It follows from (1) that there is only a finite number of distinct $r_m(x)$. Let N be an integer that is greater than the absolute value of all of the coefficients of all of the $r_m(x)$. If $p > N$ then we must have $r_p(x) = 0$ and consequently
$$f(\zeta^p) = r_p(\zeta) = 0.$$

There is only a finite number of primes p with $p \leqslant N$. Let m be the product of all these primes which do not divide n: if every prime $p \leqslant N$ divides n, let $m = 1$. Then m is prime to n. Let h be any integer relatively prime to n.

By the Chinese Remainder theorem of elementary number theory[2] there is a positive integer h' such that

$$h' \equiv h(\text{mod } n), \qquad h' \equiv 1(\text{mod } m).$$

Then $\zeta^{h'} = \zeta^h$ and every prime that divides h' is greater than N. As we have shown previously, if ξ is any primitive nth root of unity such that $f(\xi) = 0$, then $f(\xi^p) = 0$ for every prime $p > N$. Furthermore, if p does not divide n, then ξ^p is a primitive nth root of unity. Using these facts several times over, we conclude that $f(\zeta^{h'}) = 0$, and $f(\zeta^h) = 0$. Thus every primitive nth root of unity is a root of $f(x)$. This implies that $F_n(x) = f(x)$, and that $F_n(x)$ is irreducible in $Q[x]$. ∥

Since all of the nth roots of unity are powers of a primitive nth root ζ we have $K_n = Q(\zeta)$. Hence $[K_n : Q] = \phi(n)$. Since K_n is a splitting field over Q, and since Q is perfect, K_n is a Galois extension of Q. We shall now determine $G(K_n/Q)$.

Consider the set of nonzero elements of the ring $J/(n)$ of integers modulo n. The elements of $J/(n)$ are residue classes $[m]$ where $m \in J$. Set

$$G_n = \{[h] \in J/(n) \mid (h,n) = 1\}.$$

If $h \equiv h'(\text{mod } n)$ then $(h,n) = 1$ if and only if $(h',n) = 1$. Hence G_n is a well-defined subset of $J/(n)$. If $[h_1]$, $[h_2] \in G_n$ then $h_1 h_2$ is prime to n and so $[h_1][h_2] = [h_1 h_2] \in G_n$. Clearly, $[1] \in G_n$. If $[h] \in G_n$ then there is an integer h' such that $hh' \equiv 1(\text{mod } n)$. Then $[h'] \in G_n$ and $[h][h'] = [hh'] = [1]$. Thus G_n is a group which is clearly Abelian and of order $\phi(n)$.

THEOREM 7. *For any positive integer n, $G(K_n/Q) \cong G_n$.*

Proof. Let ζ be a primitive nth root of unity and let h be an integer such that $1 \leqslant h \leqslant n$ and $(h,n) = 1$. Since $K_n = Q(\zeta) = Q(\zeta^h)$ and ζ and ζ^h are both roots of $F_n(x)$, there is an element $\sigma_h \in G(K_n/Q)$ such that $\sigma_h(\zeta) = \zeta^h$. Define a mapping ψ from G_n into $G(K_n/Q)$ by $\psi([h]) = \sigma_h$. Let $[h_1]$, $[h_2] \in G_n$. Then $\sigma_{h_1}\sigma_{h_2}(\zeta) = \zeta^{h_1 h_2} = \sigma_{h_3}(\zeta)$ where $1 \leqslant h_3 \leqslant n$ and $h_3 \equiv h_1 h_2(\text{mod } n)$. Therefore $\psi([h_1])\psi([h_2]) = \psi([h_3]) = \psi([h_1][h_2])$. This shows that ψ is a homomorphism.

Let τ be any element of $G(K_n/Q)$. Then $\tau(\zeta)$ is a root of $F_n(x)$ and therefore is a primitive nth root of unity. Thus $\tau(\zeta) = \zeta^h$ for some integer h with $1 \leqslant h \leqslant n$ and $(h,n) = 1$. Then $\psi([h]) = \tau$ and ψ is *onto*. Since $G(K_n/Q)$ and G_n both have order $\phi(n)$ it follows that ψ is *one-one*, and so is an isomorphism. ∥

COROLLARY. *The nth cyclotomic field K_n is an Abelian extension of Q. All fields between Q and K_n are normal extensions of Q.*

---

[2] For the Chinese Remainder theorem, see p. 22 of [48].

Example. Let $n = 12$. Then $\phi(12) = 4$ and

$$G(K_{12}/Q) = \{\varepsilon, \sigma, \tau, \rho\},$$

where ε is the identity of the group and, if ζ is a primitive 12th root of unity,

$$\sigma(\zeta) = \zeta^5, \qquad \tau(\zeta) = \zeta^7, \qquad \rho(\zeta) = \zeta^{11}.$$

The three proper subgroups of this group are

$$H_1 = \{\varepsilon, \sigma\}, \qquad H_2 = \{\varepsilon, \tau\}, \qquad H_3 = \{\varepsilon, \rho\}.$$

If we take $\zeta = \exp(2\pi i/12)$ then

$$\sigma(\zeta^3) = \zeta^{15} = \zeta^3 \quad \text{and} \quad \zeta^3 = \exp(\pi i/2) = i,$$

and the fixed field of H_1 is $Q(i)$. Also

$$\tau(\zeta^4) = \zeta^{28} = \zeta^4 \quad \text{and} \quad \zeta^4 = \exp(2\pi i/3) = -(1 - \sqrt{3}\,i)/2.$$

Hence $\tau(\sqrt{3}\,i) = \sqrt{3}\,i$ and the fixed field of H_2 is $Q(\sqrt{3}\,i)$. Finally, $Q(\sqrt{3})$ also lies between Q and K_{12} and therefore it must be the fixed field of H_3.

5. The first cohomology group

In recent years the subject of the cohomology theory of groups has achieved a prominent position in the study of fields. We shall not develop this theory in this book but we shall give in this section a very brief introduction to some of the ideas involved, and prove certain results which will be needed later.

Let G be a group. An Abelian group A, written additively, is called a (left) *G-module* if to each $\sigma \in G$ and $a \in A$ there corresponds a unique element $\sigma(a) \in A$ and

(i) $\sigma(a + b) = \sigma(a) + \sigma(b)$,
(ii) $(\sigma\tau)(a) = \sigma(\tau(a))$,

for all $\sigma, \tau \in G$ and $a, b \in A$.

Let A be a G-module and n a non-negative integer. By an *n-cochain* of G over A we mean a function of n variables from G into A, if $n > 0$, and an element of A if $n = 0$. We denote by $C^n(G,A)$ the set of all such n-cochains. We make $C^n(G,A)$ into a group by defining

$$(f + g)(\sigma_1, \ldots, \sigma_n) = f(\sigma_1, \ldots, \sigma_n) + g(\sigma_1, \ldots, \sigma_n).$$

If $f \in C^n(G,A)$ we define an $(n + 1)$-cochain δf by

$$(\delta f)(\sigma_1, \ldots, \sigma_{n+1}) = \sigma_1(f(\sigma_2, \ldots, \sigma_{n+1}))$$
$$+ \sum_{i=1}^{n} (-1)^i f(\sigma_1, \ldots, \sigma_i\sigma_{i+1}, \ldots, \sigma_{n+1})$$
$$+ (-1)^{n+1} f(\sigma_1, \ldots, \sigma_n).$$

Then we can show that

(1) δ is a homomorphism from $C^n(G,A)$ into $C^{n+1}(G,A)$, and
(2) if $f \in C^n(G,A)$ then $\delta\delta f = 0$.

Let

$$Z^n(G,A) = \{f \in C^n(G,A) \mid \delta f = 0\},$$

and

$$B^n(G,A) = \{\delta f \mid f \in C^{n-1}(G,A)\} \qquad (n > 0),$$

$$B^0(G,A) = 0.$$

The elements of $Z^n(G,A)$ are called *n-cocycles* while the elements of $B^n(G,A)$ are called *n-coboundaries*. By (2), $B^n(G,A)$ is a subgroup of $Z^n(G,A)$. The quotient group

$$H^n(G,A) = Z^n(G,A)/B^n(G,A)$$

is called the *nth cohomology group of G over A*.

In this section we are interested in $H^1(G,A)$. Let $f \in Z^1(G,A)$, then for all $\sigma, \tau \in G$,

$$(\delta f)(\sigma,\tau) = \sigma(f(\tau)) - f(\sigma\tau) + f(\sigma) = 0$$

or

$$f(\sigma\tau) = \sigma(f(\tau)) + f(\sigma).$$

Such a mapping from G into A is called a *crossed homomorphism*. Thus

$Z^1(G,A) =$ the group of all crossed homomorphisms from G into A.

Now let $g \in B^1(G,A)$, then $g = \delta h$ for some $h \in C^0(G,A)$. Since $C^0(G,A) = A$, h is simply some element $a \in A$. Then for all $\sigma \in G$,

$$g(\sigma) = (\delta h)(\sigma) = \sigma(a) - a.$$

Thus

$B^1(G,A) =$ the group of all mappings g from G into
A such that there is an element $a \in A$
for which $g(\sigma) = \sigma(a) - a$ for all $\sigma \in G$.

Now let K be a finite Galois extension of a field k and let $G = G(K/k)$. The multiplicative group of K, denoted by K^*, and the additive group of K, denoted by K^+, are both G-modules, and so we can consider the cohomology groups $H^n(G,K^*)$ and $H^n(G,K^+)$. We shall use multiplicative notation for the groups $H^n(G,K^*)$. The following result is fundamental in the theory we are now beginning to develop.

THEOREM 8. *We have*

$$H^1(G,K^*) = 1 \qquad and \qquad H^1(G,K^+) = 0.$$

Proof. Let $f \in Z^1(G,K^*)$. We wish to show that $f \in B^1(G,K^*)$. Since f takes its values in K^* we have $f(\sigma) \neq 0$ for all $\sigma \in G$. Hence, by Theorem 1, there is an element $b \in K^*$ such that

$$\sum_{\tau \in G} f(\tau)\tau(b) = a \neq 0.$$

Then for all $\sigma \in G$,

$$\sigma(a) = \sum_{\tau \in G} \sigma(f(\tau))\sigma\tau(b).$$

Since $f \in Z^1(G,K^*)$, $f(\sigma\tau) = \sigma(f(\tau))f(\sigma)$, and therefore

$$\sigma(a) = \sum_{\tau \in G} f(\sigma)^{-1}f(\sigma\tau)\sigma\tau(b)$$
$$= f(\sigma)^{-1}\sum_{\tau \in G} f(\tau)\tau(b) = f(\sigma)^{-1}a,$$

that is, $f(\sigma) = \sigma(a^{-1})a$ for all $\sigma \in G$. Therefore, $f \in B^1(G,K^*)$.

To prove that $H^1(G,K^+) = 0$ let $f \in Z^1(G,K^+)$. Since K/k is separable it follows from Theorems 28 and 29 of Chapter 1 that there is an element $b \in K$ such that

$$\sum_{\tau \in G} \tau(b) = T_{K/k}(b) \neq 0.$$

By Corollary 1 to Theorem 26 in Chapter 1, $T_{K/k}(b) \in k$, and so

$$\sum_{\tau \in G} \tau(bT_{K/k}(b)^{-1}) = 1.$$

For our purposes we may assume from the start that

$$\sum_{\tau \in G} \tau(b) = 1.$$

Now let

$$a = \sum_{\tau \in G} f(\tau)\tau(b).$$

Since $f \in Z^1(G,K^+)$, $f(\sigma\tau) = \sigma(f(\tau)) + f(\sigma)$, and therefore, for all $\sigma \in G$,

$$\sigma(a) = \sum_{\tau \in G} \sigma(f(\tau))\sigma\tau(b)$$
$$= \sum_{\tau \in G} f(\sigma\tau)\sigma\tau(b) - f(\sigma)\sum_{\tau \in G} \sigma\tau(b)$$
$$= \sum_{\tau \in G} f(\tau)\tau(b) - f(\sigma)\sum_{\tau \in G} \tau(b)$$
$$= a - f(\sigma).$$

Thus $f(\sigma) = \sigma(-a) + a$ for all $\sigma \in G$ and $f \in B^1(G,K^+)$. ‖

That part of the next theorem which refers to the norm is known as Hilbert's Theorem 90.

THEOREM 9. *Let K be a cyclic extension of k and let σ be a generator of $G(K/k)$. If $a \in K$ and $N_{K/k}(a) = 1$ then there is an element $b \in K^*$ such that*

$a = \sigma(b)b^{-1}$. *If $a \in K$ and $T_{K/k}(a) = 0$ then there is an element $b \in K$ such that $a = \sigma(b) - b$.*

Proof. Let $a \in K$ and suppose that $N_{K/k}(a) = 1$. Define the mapping f from $G = G(K/k)$ into K^* by

$$f(1) = 1, \quad f(\sigma) = a,$$

$$f(\sigma^i) = \sigma^{i-1}(a) \cdots \sigma(a)a, \quad i = 2, \ldots, \quad n-1,$$

where $n = [K:k] = \#G$. Then $f \in Z^1(G, K^*)$. In particular,

$$\sigma^{n-1}(f(\sigma))f(\sigma^{n-1}) = \sigma^{n-1}(a)\sigma^{n-2}(a) \cdots \sigma(a)a,$$

$$= N_{K/k}(a) = 1 = f(\sigma^n).$$

By Theorem 8, $f \in B^1(G, K^*)$ and as a result there is an element $b \in K^*$ such that $\sigma(b)b^{-1} = f(\sigma) = a$.

Assume that $a \in K$ and that $T_{K/k}(a) = 0$. Define the mapping g from G into K^+ by

$$g(1) = 0, \quad g(\sigma) = a,$$

$$g(\sigma^i) = \sigma^{i-1}(a) + \cdots + \sigma(a) + a, \quad i = 2, \ldots, n-1.$$

Then $g \in Z^1(G, K^+)$, and by Theorem 8, $g \in B^1(G, K^+)$. Hence there is an element $b \in K^+$ such that $\sigma(b) - b = g(\sigma) = a$. ‖

6. Cyclic extensions

In this section we shall study the cyclic extensions of a field, and characterize them in certain cases. Let K be a cyclic extension of k with $[K:k] = n$. If char $k = p$ we write $n = mp^t$ where m is not divisible by p. If $G = G(K/k)$ then every subgroup and every quotient group of G is cyclic. Let H be the subgroup of order m of G and let F be the fixed field of H. Then $[K:F] = m$, which is prime to p, and $[F:k] = p^t$. The group $G(F/k)$ has a chain of subgroups starting with 1 and ending with $G(F/k)$, each of index p in the next one, therefore F can be obtained from k by a sequence of cyclic extensions of degree p.

Thus we may confine our study to two cases. In the first case n is prime to p if char $k = p$ or arbitrary if char $k = 0$; in the second case

$$n = \text{char } k = p.$$

In Section 4 we were concerned with complex primitive nth roots of unity. We shall now introduce the analogous concept into our discussion of an arbitrary field. An element $\zeta \in k$ is called a *primitive nth root of unity* if $\zeta^n = 1$ but $\zeta^d \neq 1$ for $1 \leq d \leq n-1$. If k contains a primitive nth root of unity ζ then it contains n distinct nth roots of unity ζ, ζ^2, ..., ζ^{n-1}, $\zeta^n = 1$, and the polynomial

$$x^n - 1 = \prod_{i=1}^{n}(x - \zeta^i)$$

has simple roots. Therefore, its derivative nx^{n-1} must not vanish. This means that either char $k = 0$ or char $k = p$ and n is not divisible by p.

THEOREM 10. *Let k contain a primitive nth root of unity ζ. Then K/k is cyclic of degree n if and only if K is the splitting field over k of an irreducible polynomial $x^n - a \in k[x]$. In this case, $K = k(b)$ where b is a root of this polynomial.*

Proof. First suppose that K/k is cyclic of degree n. Let σ be a generator of $G(K/k)$. We have $N_{K/k}(\zeta) = \zeta^n = 1$ and so by Theorem 9 there is an element $b \in K^*$ such that $\zeta = \sigma(b)b^{-1}$ or $\sigma(b) = \zeta b$. Then for $1 \leqslant i \leqslant n$ we have $\sigma^i(b) = \zeta^i b$. It follows that the $\sigma^i(b)$ are distinct and that $\sigma(b^n) = \sigma(b)^n = \zeta^n b^n = b^n$ so that $b^n = a \in k$. We have

$$x^n - a = \prod_{i=1}^{n}(x - \zeta^i b).$$

Suppose that $x^n - a = f(x)g(x)$ where $f(x)$ is a nonconstant irreducible polynomial in $k[x]$. If $\zeta^i b$ is one root of $f(x)$, then for every j we have $\sigma^{j-i}(\zeta^i b) = \zeta^j b$ and so $\zeta^j b$ is also a root of $f(x)$. Thus $x^n - a$ must itself be irreducible in $k[x]$. Thus $[k(b):k] = n$ and so $K = k(b)$ and the necessity of the stated condition is proved.

Conversely, suppose that K is a splitting field over k of an irreducible polynomial $x^n - a \in k[x]$. Let b be a root of this polynomial in K, in which case the k-isomorphism from $k(b)$ onto $k(\zeta b)$ which maps b onto ζb can be extended to a k-automorphism σ of K. For $1 \leqslant i \leqslant n$, $\sigma^i(b) = \zeta^i b$, therefore $\sigma, \sigma^2, \ldots, \sigma^n$ are distinct elements of $G(K/k)$. Since the roots of $x^n - a$ are $\zeta^i b$, $1 \leqslant i \leqslant n$, we have $K = k(b)$ so that $[K:k] = n \geqslant \#G(K/k) \geqslant n$. Hence $G(K/k)$ has order n from which it follows that K/k is separable, and so Galois, and $G(K/k)$ is cyclic with σ as a generator. ∥

THEOREM 11. *Let char $k = p$. Then K/k is cyclic of degree p if and only if K is a splitting field over k of an irreducible polynomial $x^p - x - a \in k[x]$. In this case, $K = k(b)$ where b is a root of this polynomial.*

Proof. First, suppose that K/k is cyclic of degree p and let σ be a generator of $G(K/k)$. Since $T_{K/k}(1) = p \cdot 1 = 0$ there is, by Theorem 9, an element $b \in K$ such that $1 = \sigma(b) - b$ or $\sigma(b) = b + 1$. Define the polynomial $\mathcal{P}(x)$ by $\mathcal{P}(x) = x^p - x$. Then $\sigma(\mathcal{P}(b)) = \sigma(b^p - b) = \sigma(b)^p - \sigma(b) = (b + 1)^p - (b + 1) = b^p - b = \mathcal{P}(b)$, so that $\mathcal{P}(b) = a \in k$. Thus

b is a root of $x^p - x - a$. If $1 \leqslant i \leqslant p$ then $\sigma^i(b) = b + i$ and $(b + i)^p - (b + i) - a = b^p - b - a = 0$ so that

$$x^p - x - a = \prod_{i=1}^{p}(x - b - i) = \prod_{i=1}^{p}(x - \sigma^i(b)).$$

However, this implies that $x^p - x - a$ is irreducible in $k[x]$, so that $[k(b):k] = p$. Therefore $K = k(b)$.

Conversely, assume that K is a splitting field over k of an irreducible polynomial $x^p - x - a \in k[x]$. Let b be a root of this polynomial in K. Then $b + 1$ is also a root of this polynomial in K and there is a k-automorphism σ of K such that $\sigma(b) = b + 1$. Then the roots of $x^p - x - a$ in K are $\sigma^i(b) = b + i$ for $1 \leqslant i \leqslant p$, and they are distinct. Hence $\sigma, \sigma^2, \ldots, \sigma^p$ are distinct elements of $G(K/k)$. Furthermore, $K = k(b)$ so that $[K:k] = p \geqslant \#G(K/k) \geqslant p$. Thus $G(K/k)$ has order p and K/k is cyclic of degree p with σ as a generator of $G(K/k)$. ∥

7. Multiplicative Kummer theory

Whereas in the last section we have studied the cyclic extensions of k, in this section and the next one we shall study the finite Abelian extensions of k.

Let G be a finite Abelian group of order n. Then G has a *basis* $g_1, \ldots, g_r$ such that every element $g \in G$ can be written uniquely in the form

$$g = g_1^{\alpha_1} \cdots g_r^{\alpha_r}, \qquad \alpha_1, \ldots, \qquad \alpha_r \text{ integers}, \qquad 1 \leqslant \alpha_i \leqslant \gamma_i,$$

where γ_i is the order of g_i. Note that γ_i divides n. Let k be a field that contains a primitive nth root of unity. A homomorphism χ from G into k^* is called a *character* of G in k. For $i = 1, \ldots, r$,

$$\chi(g_i)^{\gamma_i} = \chi(g_i^{\gamma_i}) = \chi(1) = 1,$$

and so $\chi(g_i)$ is a γ_ith root of unity; it is therefore an nth root of unity. Let $G^\wedge$ be the set of all characters of G in k. If $\chi_1, \chi_2 \in G^\wedge$, we define their product $\chi_1\chi_2$ by $(\chi_1\chi_2)(g) = \chi_1(g)\chi_2(g)$. With respect to this operation $G^\wedge$ forms a group which is called the *dual group* of G.

THEOREM 12. *If G is a finite Abelian group then $G \cong G^\wedge$.*

Proof. We shall find a basis $\chi_1, \ldots, \chi_r$ of $G^\wedge$ such that the order of χ_i is γ_i. Since k has a primitive nth root of unity and since γ_i divides n, k has a primitive γ_ith root of unity, namely, the (n/γ_i)th power of a primitive nth root of unity. Let ζ_i be a primitive γ_ith root of unity in k and define χ_i by

$$\chi_i(g_j) = \begin{cases} \zeta_i & \text{if} \quad j = i \\ 1 & \text{if} \quad j \neq i. \end{cases}$$

If $g = g_1^{\alpha_1} \cdots g_r^{\alpha_r}$ then $\chi_i(g) = \zeta_i^{\alpha_i}$. If $\chi \in G\hat{}$ then for each i, $\chi(g_i) = \zeta_i^{\beta_i}$ for some integer β_i with $1 \leqslant \beta_i \leqslant \gamma_i$ and we have

$$\chi(g) = \chi(g_1^{\alpha_1} \cdots g_r^{\alpha_r}) = \chi(g_1)^{\alpha_1} \cdots \chi(g_r)^{\alpha_r}$$
$$= \zeta_1^{\alpha_1\beta_1} \cdots \zeta_r^{\alpha_r\beta_r} = \chi_1(g)^{\beta_1} \cdots \chi_r(g)^{\beta_r}$$

so that

$$\chi = \chi_1^{\beta_1} \cdots \chi_r^{\beta_r}.$$

Note that the order of χ_i is γ_i.

Now suppose that for non-negative integers $\beta_1, \ldots, \beta_r$, $\chi_1^{\beta_1} \cdots \chi_r^{\beta_r}$ is the identity element of $G\hat{}$. Then for $i = 1, \ldots, r$,

$$1 = (\chi_1^{\beta_1} \cdots \chi_r^{\beta_r})(g_i) = \zeta_i^{\beta_i},$$

and since ζ_i is a primitive γ_ith root of unity this implies that γ_i divides β_i. Thus $\chi_1, \ldots, \chi_r$ form a basis of $G\hat{}$. ‖

Let G and H be Abelian groups and A a finite cyclic group. By a *pairing* of G and H into A we mean a mapping ϕ from $G \times H$ into A such that

$$\phi(g_1g_2,h) = \phi(g_1,h)\phi(g_2,h)$$

and

$$\phi(g,h_1h_2) = \phi(g,h_1)\phi(g,h_2)$$

for all g, g_1, $g_2 \in G$ and h, h_1, $h_2 \in H$. We define subgroups G' of G and H' of H by

$$G' = \{g \in G \mid \phi(g,h) = 1 \text{ for all } h \in H\},$$
$$H' = \{h \in H \mid \phi(g,h) = 1 \text{ for all } g \in G\}.$$

THEOREM 13. *If H/H' is finite then $G/G' \cong H/H'$.*

Proof. We may assume that A is the cyclic group of mth roots of unity in the field of complex numbers, where m is the order of A. If $g \in G$ we define $\chi_g \in H\hat{}$ by $\chi_g(h) = \phi(g,h)$. For $h \in H'$, $\chi_g(h) = 1$ and therefore we may regard χ_g as belonging to $(H/H')\hat{}$. For $g_1, g_2 \in G$ we have

$$\chi_{g_1g_2}(h) = \phi(g_1g_2,h) = \phi(g_1,h)\phi(g_2,h)$$
$$= (\chi_{g_1}\chi_{g_2})(h),$$

and so the mapping $g \to \chi_g$ is a homomorphism from G into $(H/H')\hat{}$. If g is in the kernel of this homomorphism then for all $h \in H$, $\chi_g(h) = \phi(g,h) = 1$ and $g \in G'$. Conversely, any $g \in G'$ is in the kernel of this homomorphism. Hence G/G' is isomorphic to a subgroup of $(H/H')\hat{}$ which in turn is isomorphic to H/H' by Theorem 12. Hence G/G' is isomorphic to a subgroup of H/H'. As such, G/G' is finite, and by a similar argument, H/H' is isomorphic to a subgroup of G/G'. Since both G/G' and H/H' are finite this implies that $G/G' \cong H/H'$. ‖

Assume that the field k contains a primitive nth root of unity. As we have noted previously this implies that either char $k = 0$ or char $k = p$ and n is not divisible by p. Let C be an algebraic closure of k. Let $k^{*n} = \{a^n \mid a \in k^*\}$—this is a subgroup of k^*. In fact, if A is any subgroup of C^* we set $A^n = \{a^n \mid a \in A\}$. If S is any subset of k we set $S^{1/n} = \{a \in C \mid a^n \in S\}$. If K is a finite Abelian extension of k we say that K has *exponent* n if the order of each element of $G(K/k)$ divides n. We can now state and prove the principal result of this section.

THEOREM 14. *Let k be a field that contains a primitive nth root of unity. Then there is a one-one correspondence between the subgroups H of k^* which contain k^{*n} and for which the index $(H:k^{*n})$ is finite and the finite Abelian extensions of exponent n of k in C. The correspondence is*

$$H \leftrightarrow K_H = k(H^{1/n}).$$

Furthermore, $G(K_H/k) \cong H/k^{*n}$ so that $[K_H:k] = (H:k^{*n})$.

Proof. The proof will be given in several steps.

Step 1.   Let K be a finite Abelian extension of exponent n of k in C and let $G = G(K/k)$. Let

$$A = A_K = \{a \in K^* \mid a^n \in k^*\}.$$

It is clear that A is a subgroup of K^*. Let χ be a character of G in k. Then for all σ, $\tau \in G$, $\sigma(\chi(\tau)) = \chi(\tau)$ and so $\chi(\sigma\tau) = \chi(\sigma)\chi(\tau) = \sigma(\chi(\tau))\chi(\sigma)$, that is, $\chi \in Z^1(G,K^*)$. Then by Theorem 8, $\chi \in B^1(G,K^*)$ so that there is an element $a \in K^*$ such that $\chi(\sigma) = a\sigma(a)^{-1}$ for all $\sigma \in G$. Since $a^n\sigma(a^n)^{-1} = \chi(\sigma)^n = \chi(\sigma^n) = \chi(1) = 1$ we have $\sigma(a^n) = a^n$ for all $\sigma \in G$ so that $a^n \in k^*$, that is, $a \in A$. On the other hand, let $a \in A$ and set $\chi(\sigma) = a\sigma(a)^{-1}$ for all $\sigma \in G$. Then $\chi(\sigma)^n = a^n\sigma(a^n)^{-1} = 1$, so that $\chi(\sigma)$ is an nth root of unity, and therefore in k, and as we show in the next paragraph, $\chi(\sigma\tau) = \chi(\sigma)\chi(\tau)$. Therefore, χ is a character of G in k.

Now consider the mapping ϕ from $A \times G$ into the group of nth roots of unity in k^* given by

$$\phi(a,\sigma) = a\sigma(a)^{-1}.$$

We have

$$\phi(ab,\sigma) = ab\sigma(ab)^{-1} = a\sigma(a)^{-1}b\sigma(b)^{-1}$$
$$= \phi(a,\sigma)\phi(b,\sigma)$$

and

$$\phi(a,\sigma\tau) = a(\sigma\tau)(a)^{-1} = a\sigma(a)^{-1}\sigma(a)(\sigma\tau)(a)^{-1}$$
$$= a\sigma(a)^{-1}\sigma(a\tau(a)^{-1})$$
$$= \phi(a,\sigma)\phi(a,\tau),$$

since $\phi(a,\tau) \in k$. Thus ϕ is a pairing of A and G into a finite cyclic group.

Set
$$G' = \{\sigma \in G \mid \phi(a,\sigma) = 1 \text{ for all } a \in A\}$$
and
$$A' = \{a \in A \mid \phi(a,\sigma) = 1 \text{ for all } \sigma \in G\}.$$

Since G/G' is finite we have $G/G' \cong A/A'$, by Theorem 13.

We have
$$\begin{aligned}
A' &= \{a \in A \mid a\sigma(a)^{-1} = 1 \text{ for all } \sigma \in G\} \\
&= \{a \in A \mid \sigma(a) = a \text{ for all } \sigma \in G\} \\
&= A \cap k^* = k^*.
\end{aligned}$$

Also
$$G' = \{\sigma \in G \mid \sigma(a) = a \text{ for all } a \in A\}.$$

Thus $\sigma \in G'$ if and only if $\chi(\sigma) = 1$ for all $\chi \in G^\wedge$. However, it follows from the proof of Theorem 12 that, given any $\sigma \in G$ there is a $\chi \in G^\wedge$ such that $\chi(\sigma) \neq 1$, unless $\sigma = 1$. Therefore, $G' = 1$ and $G/G' \cong G$. Consequently, we have shown that
$$G \cong A/k^*.$$

The mapping $a \to a^n$ is a homomorphism from A onto A^n which maps k^* onto k^{*n}. Thus it induces a homomorphism $ak^* \to a^n k^{*n}$ from A/k^* onto A^n/k^{*n}. If $ak^* \to k^{*n}$ then $a^n \in k^{*n}$ and $a^n = b^n$ for some $b \in k^*$. But then ab^{-1} is an nth root of unity and so lies in k^*, which implies that $a \in k^*$. It follows that $A/k^* \cong A^n/k^{*n}$, and
$$G \cong A^n/k^{*n}.$$

Since G is finite, $(A^n : k^{*n})$ is finite.

Thus there is a mapping $K \to A_K{}^n$, from the set of finite Abelian extensions of exponent n of k in C into the set of subgroups H of k^* which contain k^{*n} such that $(H : k^{*n})$ is finite.

Step 2. Let H be a subgroup of k^* which contains k^{*n} such that $(H : k^{*n})$ is finite. Let $K = K_H = k(H^{1/n})$. Let $a_1, \ldots, a_r$ be elements of H which form a complete set of representatives of the cosets of k^{*n} in H. Let F be the splitting field of $(x^n - a_1) \cdots (x^n - a_r)$ over k in C. Then $F \subseteq K$. Let $b \in H^{1/n}$. Then $b^n \in H$ and there is an element $c \in k^*$ such that $b^n = a_i c^n$ for some i with $1 \leqslant i \leqslant r$. Hence $b = d_i c$ where d_i is a root of $x^n - a_i$ in C. Thus $b \in F$. Consequently, we have $H^{1/n} \subseteq F$ and therefore $K \subseteq F$. As a result $K = F$. If d_i is one root of $x^n - a_i$ in C then the other roots of this polynomial in C are $\zeta d_i, \ldots, \zeta^{n-1} d_i$, where ζ is a primitive nth root of unity in k. Hence K is both normal and separable over k and $K = k(d_1, \ldots, d_r)$.

Let $G = G(K/k)$. We shall show that G is Abelian and that the order of each element of G divides n. In order to show that G is Abelian it is sufficient to show that if $\sigma, \tau \in G$ and $a \in H^{1/n}$, then $\sigma\tau(a) = \tau\sigma(a)$. We have

$a^n \in H \subseteq k^*$ and a is a root of $x^n - a^n$. Since $\sigma(a)$ is another root of this polynomial we have $\sigma(a) = \zeta^r a$ for some integer r and primitive nth root of unity ζ in k. Similarly, $\tau(a) = \zeta^s a$. Then $\sigma\tau(a) = \sigma(\zeta^s a) = \zeta^{s+r} a = \tau(\zeta^r a) = \tau\sigma(a)$. Furthermore, $\sigma^n(a) = \zeta^{rn} a = a$ for all $a \in H$ and therefore for all $a \in K$. Hence $\sigma^n = 1$ so that the order of σ divides n.

Thus there is a mapping $H \to K_H$ from the set of subgroups H of k^* which contain k^{*n}, and such that $(H : k^{*n})$ is finite, into the set of finite Abelian extensions of exponent n of k in C.

Step 3. We shall now show that the mapping described in Step 2 is *one-one* and *onto*.

Onto. Let K be a finite Abelian extension of exponent n of k in C. Let $A = A_K$ and $H = A^n$. Then $A \subseteq H^{1/n}$, and since every element of H has one nth root in K and k contains a primitive nth root of unity, we have $H^{1/n} \subseteq K$. Hence $K_H \subseteq K$. By Step 2, K_H is a finite Abelian extension of exponent n of k in C, and so if $A_1 = A_{K_H}$, we have $G(K_H/k) \cong A_1/k^*$, by Step 1. However, $A \subseteq H^{1/n} \subseteq A_1 \subseteq A$ so that $A_1 = A$. Thus, $[K:k] = \#G(K/k) = (A:k^*) = \#G(K_H/k) = [K_H:k]$. This implies that $K = K_H$.

One-one. It is sufficient to show that if H is a subgroup of k^* which contains k^{*n}, and such that $(H : k^{*n})$ is finite and if $K = K_H$, then $H = A_K^n$. Using the same argument as in the preceding paragraph, it follows that the mapping $N \to K_N$ maps the set of all subgroups of k^* such that $k^{*n} \subseteq N \subseteq A_K^n$ *onto* the set of all finite Abelian extensions of exponent n of k which are contained in K: this is precisely the set of all fields between k and K. However, the number of such subgroups is equal to the number of subgroups of $G(K/k)$, since $G(K/k) \cong A_K^n/k^{*n}$, which is equal to the number of fields between k and K by the fundamental theorem of Galois theory (Theorem 5). Therefore, the mapping $N \to K_N$ is one-one. Since H and A_K^n have the same image under this mapping they must be equal. ‖

DEFINITION. Let k be a field that contains a primitive nth root of unity. A finite Abelian extension of exponent n of k is called a *Kummer extension* of k. Thus the Kummer extensions of k are the splitting fields over k of polynomials of the form $(x^n - a_1) \cdots (x^n - a_r)$ where $a_1, \ldots, a_r \in k$.

8. Additive Kummer theory

Let k be a field of characteristic p and let C be an algebraic closure of k. If B is any subgroup of C^+ we set

$$\mathscr{P}B = \{\mathscr{P}(b) \mid b \in B\},$$

where $\mathscr{P}(x) = x^p - x$. Since $\mathscr{P}(b_1 + b_2) = \mathscr{P}(b_1) + \mathscr{P}(b_2)$, $\mathscr{P}B$ is a subgroup of C^+. If S is any subset of k we set

$$\mathscr{P}^{-1}S = \{a \in C \mid \mathscr{P}(a) \in S\}.$$

In particular, if $a \in k$ then $\mathscr{P}^{-1}a$ is the set of all solutions in C of the equation $x^p - x - a = 0$.

THEOREM 15. *If k is a field of characteristic p then there is a one-one correspondence between the finite Abelian extensions of exponent p of k in C and the subgroups H of k^+ which contain $\mathscr{P}k^+$ and for which the index $(H : \mathscr{P}k^+)$ is finite. The correspondence is*

$$H \leftrightarrow K_H = k(\mathscr{P}^{-1}H).$$

Furthermore, $G(K/k) \cong H/\mathscr{P}k^+$.

Proof. Let K be a finite Abelian extension of exponent p of k in C. Let $G = G(K/k)$ and let $A = \{a \in K^+ \mid \mathscr{P}(a) \in k\}$. If $a, b \in A$ then $\mathscr{P}(a - b) = \mathscr{P}(a) - \mathscr{P}(b) \in k$ so that A is a subgroup of K^+. Let χ be a homomorphism from G into $GF(p)^+$, the additive group of the prime field of k. Then for all $\sigma, \tau \in G$ we have $\chi(\sigma\tau) = \chi(\sigma) + \chi(\tau) = \sigma(\chi(\tau)) + \chi(\sigma)$. Hence $\chi \in Z^1(G,K^+)$. Therefore, by Theorem 8, $\chi \in B^1(G,K^+)$, and so there is an element $a \in K$ such that $\chi(\sigma) = a - \sigma(a)$ for all $\sigma \in G$. Then, for all $\sigma \in G$, $\mathscr{P}(a) - \sigma(\mathscr{P}(a)) = a^p - a - \sigma(a^p) + \sigma(a) = (a - \sigma(a))^p - (a - \sigma(a)) = \chi(\sigma)^p - \chi(\sigma) = 0$ since $\chi(\sigma)$ lies in the prime field of k. Thus $\mathscr{P}(a) \in k$ and so $a \in A$. On the other hand, let $a \in A$ and define χ by $\chi(\sigma) = a - \sigma(a)$. Then $\mathscr{P}(\chi(\sigma)) = (a - \sigma(a))^p - (a - \sigma(a)) = (a^p - a) - (\sigma(a)^p - \sigma(a)) = \mathscr{P}(a) - \sigma(\mathscr{P}(a)) = 0$ and so $\chi(\sigma)$ is a root of $\mathscr{P}(x)$. However, $\mathscr{P}(x)$ has at most p roots in k and each of the p elements of the prime field of k is a root of $\mathscr{P}(x)$, so that we conclude that $\chi(\sigma)$ lies in the prime field of k. Moreover, χ is a homomorphism from G into $GF(p)^+$, since we have for $\sigma, \tau \in G$,

$$\chi(\sigma\tau) = a - \sigma\tau(a)$$
$$= (a - \sigma(a)) + \sigma(a - \tau(a))$$
$$= \chi(\sigma) + \sigma(\chi(\tau)) = \chi(\sigma) + \chi(\tau).$$

Define the mapping ϕ from $A \times G$ into the additive group of the prime field of k by

$$\phi(a,\sigma) = a - \sigma(a).$$

This is a pairing of A and G into $GF(p)^+$ and it follows from this that $G \cong A/k^+$. The mapping $a \to \mathscr{P}(a)$ from A onto $\mathscr{P}A$ induces an isomorphism $A/k^+ \cong \mathscr{P}A/\mathscr{P}k^+$, and so we have $G \cong \mathscr{P}A/\mathscr{P}k^+$. The proof of these statements, in fact, the entire proof, is quite similar to the proof of Theorem 14 and is left to the reader as an exercise.

9. Solutions of polynomial equations by radicals

Let k be a field and let $f(x)$ be a nonconstant polynomial in $k[x]$. We shall consider in this section the problem of solving the equation $f(x) = 0$. If $f(x) = x^2 + bx + c$, we know that if char $k \neq 2$ then the solutions of $f(x) = 0$ are given by $(-b \pm \sqrt{b^2 - 4c})/2$ so that this equation either has solutions in k or solutions in $k(\sqrt{b^2 - 4c})$. Thus $f(x) = 0$ can always be solved in a field obtained by adjoining a radical to k. We wish to determine when an arbitrary polynomial equation $f(x) = 0$ can be solved, if not in k itself, at least in some field obtained from k by the successive adjunction of radicals. In order to simplify our presentation a little we shall assume throughout this section that char $k = 0$.

DEFINITION. We say that $f(x) = 0$ is *solvable by radicals* if there are fields $k = F_0 \subseteq F_1 \subseteq \cdots \subseteq F_r$ such that for $i = 1, \ldots, r$, $F_i = F_{i-1}(a_i)$ where a_i is a root of $x^{n_i} - b_i$, $b_i \in F_{i-1}$, and such that F_r contains a splitting field of $f(x)$ over k. Such a sequence of fields is called a *radical tower* over k. If F_r/k is normal then it is called a *normal radical tower* over k.

THEOREM 16. *If $k = F_0 \subseteq F_1 \subseteq \cdots \subseteq F_r$ is a radical tower over k then there is a normal radical tower $k = K_0 \subseteq K_1 \subseteq \cdots \subseteq K_s$ over k such that $F_r \subseteq K_s$.*

Proof. We may assume that all of the fields with which we are concerned are subfields of some fixed algebraic closure C of k. Since $k = F_0 \subseteq F_1 \subseteq \cdots \subseteq F_r$ is a radical tower we have, for $i = 1, \ldots, r$, $F_i = F_{i-1}(a_i)$ where a_i is a root of $x^{n_i} - b_i$, $b_i \in F_{i-1}$. For each i let ζ_i be a primitive n_ith root of unity in C and let $L_1 = k(\zeta_1)$, $L_2 = L_1(\zeta_2), \ldots, L_r = L_{r-1}(\zeta_r)$. Then set $L_{r+1} = L_r(a_1), \ldots, L_{2r} = L_{2r-1}(a_r)$. Note that for $j = 1, \ldots, r$, $b_j \in F_{j-1} \subseteq L_{r+j-1}$. For each $i = 1, \ldots, 2r$, either $L_i = L_{i-1}$ or $L_i = L_{i-1}(a_i')$ where a_i' is a root of a polynomial of the form $x^{m_i} - c_i$, $c_i \in L_{i-1}$. For $i = 1, \ldots, r$, L_i/L_{i-1} is Abelian (see Exercise 29) and for $i = r+1, \ldots, 2r$, L_i/L_{i-1} is cyclic by Theorem 10. Thus we may assume that in our original radical tower over k, F_i/F_{i-1} is Abelian for $i = 1, \ldots, r$.

We now prove the theorem by induction on r. Since $F_0 = k$ the result is true when $r = 0$. Now suppose that $r > 0$ and that there is a normal radical tower over k whose last term, K_t, contains F_{r-1}. Since K_t/k is normal and separable we have $K_t = k(c)$ and every conjugate of c in C is also in K_t. Let ζ be a primitive n_rth root of unity in C and consider $K_t(\zeta) = k(c, \zeta)$. Every conjugate of c and every conjugate of ζ in C is also in $K_t(\zeta)$ and so $K_t(\zeta)$ is a splitting field over k of the polynomial $(x^{n_r} - 1) \operatorname{Irr}(k, c)$, which is in $k[x]$. Thus $K_t(\zeta)/k$ is normal. Since $K_t(\zeta)$ is obtained from K_t by adjoining a radical we may assume from the start that $\zeta \in K_t$.

Recall that $F_r = F_{r-1}(a_r)$ where a_r is a root of $x^{n_r} - b_r$, $b_r \in F_{r-1}$. Let $b_r^{(1)} = b_r$, $b_r^{(2)}, \ldots, b_r^{(m)}$ be the k-conjugates of b_r in C and let

$$f(x) = \prod_{i=1}^{m} (x^{n_r} - b_r^{(i)}).$$

Then $f(x) \in k[x]$ and we let K_s be a splitting field of $f(x)$ over K_t which contains $K_t(a_r)$. Then K_s is a splitting field over k of $f(x)$ Irr $(k,c) \in k[x]$ and so K_t/k is normal. For $i = 1, \ldots, m$, let $a_r^{(i)}$ be a root of $x^{n_r} - b_r^{(i)}$ in K_s. Let $K_{t+1} = K_t(a_r^{(1)}), \ldots, K_{t+m} = K_{t+m-1}(a_r^{(m)})$. Then $K_{t+m} = K_s$ and so we have obtained a radical tower over k whose last term, K_s, is normal over k and contains F_r. ‖

Let $f(x) \in k[x]$ and let K be a splitting field of $f(x)$ over k. The Galois group $G(K/k)$ will be called the Galois group of $f(x)$. The principal result of this section is that $f(x) = 0$ is solvable by radicals if and only if the Galois group of $f(x)$ belongs to a certain class of groups which we now proceed to introduce.

DEFINITION. A finite group G is said to be *solvable* if there is a chain of subgroups

$$G = G_0 \supseteq G_1 \supseteq \cdots \supseteq G_n = 1$$

such that for $i = 1, \ldots, n$, G_i is a normal subgroup of G_{i-1} and G_{i-1}/G_i is Abelian.

LEMMA. *Let G be a finite Abelian group. Then there is a chain of subgroups*

$$G = G_0 \supseteq G_1 \supseteq \cdots \supseteq G_n = 1$$

such that for $i = 1, \ldots, n$, G_{i-1}/G_i is cyclic.

Proof. By induction on the order of G. The result is true when G has only one element and so we may assume that G is of an order greater than one and that the result is true for all Abelian groups of order smaller than the order of G. If G is cyclic there is nothing to prove. Otherwise we choose an arbitrary element σ of G with $\sigma \neq 1$. Let H be the cyclic subgroup of G generated by σ. Then $H \neq 1$ and so G/H is of order smaller than the order of G. By our induction assumption there is a chain of subgroups $G = G_0 \supseteq G_1 \supseteq \cdots \supseteq G_m = H$ such that for $i = 1, \ldots, m$,

$$\frac{G_{i-1}/H}{G_i/H} \cong G_{i-1}/G_i$$

is cyclic. Since H is cyclic the lemma follows. ‖

PROPOSITION 1. *A finite group G is solvable if and only if there is a chain of subgroups*

$$G = G_0 \supseteq G_1 \supseteq \cdots \supseteq G_n = 1$$

such that for $i = 1, \ldots, n$, G_i is a normal subgroup of G_{i-1} and G_{i-1}/G_i is cyclic.

Proof. The condition is sufficient by the definition of solvable group and necessary by virtue of the lemma. ‖

PROPOSITION 2. *Let G be a finite solvable group and H a subgroup of G. Then H is solvable. If H is a normal subgroup of G then G/H is solvable. Conversely, if G is a finite group and if there is a normal subgroup H of G such that both H and G/H are solvable then G is solvable.*

Proof. First, assume that G is solvable and let $G = G_0 \supseteq G_1 \supseteq \cdots \supseteq G_n = 1$ be a chain of subgroups of G such that for $i = 1, \ldots, n$, G_i is normal in G_{i-1} and G_{i-1}/G_i is Abelian. Let H be a subgroup of G and let $H_i = H \cap G_i$. For each i, H_i is normal in H_{i-1} since $H_i = H_{i-1} \cap G_i$. Since G_i is normal in G_{i-1}, $H_{i-1}G_i$ is a subgroup of G_{i-1}. Then $H_{i-1}/H_i = H_{i-1}/H_{i-1} \cap G_i \cong H_{i-1}G_i/G_i$ which is a subgroup of G_{i-1}/G_i. Since this last group is Abelian, H_{i-1}/H_i is Abelian. Thus H is solvable. If H is normal in G then we have $G_0/H \supseteq G_1H/H \supseteq \cdots \supseteq G_nH/H$ and for $i = 1, \ldots, n$, the natural homomorphism from G onto G/H induces a homomorphism from G_{i-1}/G_i onto $(G_{i-1}H/H)/(G_iH/H)$. Since a homomorphic image of an Abelian group is Abelian it follows that G/H is solvable.

On the other hand, let H be a normal subgroup of G such that both H and G/H are solvable. Then there is a chain of subgroups $H = H_0 \supseteq H_1 \supseteq \cdots \supseteq H_r = 1$ such that for $i = 1, \ldots, r$, H_i is normal in H_{i-1} and H_{i-1}/H_i is Abelian. Also there is a chain of subgroups $G = G_0 \supseteq G_1 \supseteq \cdots \supseteq G_s = H$ such that for $i = 1, \ldots, s$, G_i is normal in G_{i-1} and $G_{i-1}/G_i \cong (G_{i-1}/H)/(G_i/H)$ is Abelian. Thus G is solvable. ‖

THEOREM 17. *Let $f(x)$ be a nonconstant polynomial in $k[x]$. Then $f(x) = 0$ is solvable by radicals if and only if the Galois group of $f(x)$ is solvable.*

Proof. Let K be a splitting field over k of $f(x)$ and assume that $G = G(K/k)$ is solvable. We shall show that $f(x) = 0$ is solvable by radicals. Let $n = [K:k] = \#G(K/k)$.

Assume first that k contains a primitive nth root of unity. Then k contains a primitive mth root of unity for all positive integers m that divide n. Let $G = G_0 \supseteq G_1 \supseteq \cdots \supseteq G_r = 1$ be a chain of subgroups of G such that for $i = 1, \ldots, r$, G_i is a normal subgroup of G_{i-1} and G_{i-1}/G_i is cyclic. Let F_i be the fixed field of G_i. Then $k = F_0 \subseteq F_1 \subseteq \cdots \subseteq F_r = K$. If $n_i = [F_i:F_{i-1}]$ then n_i divides n and so F_{i-1} contains a primitive n_ith root of unity. Furthermore, F_i/F_{i-1} is cyclic since $G(F_i/F_{i-1}) \cong G_{i-1}/G_i$. Then by Theorem 10, $F_i = F_{i-1}(a_i)$ where a_i is a root of $x^{n_i} - b_i$, $b_i \in F_{i-1}$. Thus $k = F_0 \subseteq F_1 \subseteq \cdots \subseteq F_r = K$ is a radical tower over k and F_r is a splitting field of $f(x)$ over k, so $f(x) = 0$ is solvable by radicals.

Now we drop the assumption that k contains a primitive nth root of unity. Let C be an algebraic closure of k which contains K and let ζ be a primitive nth root of unity in C. Let $L = k(\zeta)$ and $K_1 = KL$. Then K_1 is a splitting field of $f(x)$ over L and $G(K_1/L)$ is isomorphic to a subgroup of G by Theorem 6. Hence, by Proposition 2, $G(K_1/L)$ is solvable. If $m = [K_1 : L]$ then m divides n and therefore L contains a primitive mth root of unity. Then by the first part of the proof there is a radical tower $L = F_0 \subseteq F_1 \subseteq \cdots \subseteq F_r = K_1$. But then $k \subseteq k(\zeta) = L = F_0 \subseteq F_1 \subseteq \cdots \subseteq F_r = K_1$ is a radical tower over k and since $K \subseteq K_1$ it follows that $f(x) = 0$ is solvable by radicals.

Conversely, suppose that $f(x) = 0$ is solvable by radicals. By Theorem 16 there is a normal radical tower $k = K_0 \subseteq K_1 \subseteq \cdots \subseteq K_r$ over k such that K_r contains a splitting field K of $f(x)$ over k. We have, for $i = 1, \ldots, r$, $K_i = K_{i-1}(a_i)$ where a_i is a root of $x^{n_i} - b_i$, $b_i \in K_{i-1}$. Let $n = n_1 \cdots n_r$ and let ζ be a primitive nth root of unity in some algebraic closure of k which contains K_r. For $i = 1, \ldots, r$ let $K_i' = K_i(\zeta)$. Then $k(\zeta) = K_0' \subseteq K_1' \subseteq \cdots \subseteq K_r'$ is a normal radical tower over $k(\zeta)$. We have $K_i' = K_{i-1}(\zeta, a_i) = K_{i-1}'(a_i)$, and K_{i-1}' contains a primitive n_ith root of unity, and so by Theorem 10, K_i'/K_{i-1}' is cyclic. Let $H_i = G(K_r'/K_i')$, then $G(K_r'/k(\zeta)) = H_0 \supseteq H_1 \supseteq \cdots \supseteq H_r = 1$ and $H_{i-1}/H_i = G(K_r'/K_{i-1}')/G(K_r'/K_i') \cong G(K_i'/K_{i-1}')$, which is cyclic. Thus $G(K_r'/k(\zeta))$ is solvable. Since $K_r' = K_r k(\zeta)$ we have, by Theorem 6,

$$G(K_r'/k(\zeta)) \cong G(K_r/K_r \cap k(\zeta)),$$

so that the group on the right is solvable. Furthermore, by Exercise 29, $k(\zeta)$ is an Abelian extension of k (we may even have $k(\zeta) = k$), and consequently $K_r \cap k(\zeta)$ is an Abelian extension of k. Thus $G(K_r \cap k(\zeta)/k)$ is solvable. Hence $G(K_r/k)$ has a solvable normal subgroup $G(K_r/K_r \cap k(\zeta))$ such that the factor group $G(K_r/k)/G(K_r/K_r \cap k(\zeta)) \cong G(K_r \cap k(\zeta)/k)$ is solvable. Then, by Proposition 2, $G(K_r/k)$ is solvable and its factor group $G(K/k)$ is solvable. ‖

Assume that $f(x)$ has degree 2, 3 or 4. Then the Galois group of $f(x)$ is a subgroup of S_2, S_3, or S_4, respectively, where S_n is the symmetric group on n letters. Each of these groups is solvable. Let A_n be the alternating subgroup of S_n. The S_2 has order 2, and therefore is Abelian: $S_3 \supset A_3 \supset 1$ and S_3/A_3 has order 2 and A_3 has order 3; $S_4 \supset A_4 \supset H \supset 1$ where H is the noncyclic group of order 4, so that S_4/A_4 has order 2 and A_4/H and H have orders 3 and 4, respectively. Thus, if $f(x)$ has degree 2, 3, or 4 then $f(x) = 0$ is solvable by radicals.

Assume that $n \geqslant 5$. We shall show that there is a field k of characteristic 0 and an irreducible polynomial $f(x) \in k[x]$ with $\deg f(x) = n$ such that $f(x) = 0$ is not solvable by radicals. Let F be an arbitrary field of characteristic 0 and let $k = F(y_1, \ldots, y_n)$ where $y_1, \ldots, y_n$ are independent indeterminates over F. Let

$$f(x) = x^n + y_1 x^{n-1} + \cdots + y_n$$

and let K be a splitting field of $f(x)$ over k.

THEOREM 18. *The polynomial $f(x)$ is irreducible in $k[x]$ and $[K:k] = n!$. Thus $G(K/k) \cong S_n$.*

Proof. All we need to do is to show that $[K:k] = n!$. Let $a_1, \ldots, a_n$ be the roots of $f(x)$ in K. Then $K = k(a_1, \ldots, a_n)$. In fact, since for each i, y_i is a symmetric function of $a_1, \ldots, a_n$, we have $K = F(a_1, \ldots, a_n)$. Now let $z_1, \ldots, z_n$ be independent indeterminates over F and let $K_1 = F(z_1, \ldots, z_n)$. Let $b_1, \ldots, b_n$ be the elementary symmetric functions in $z_1, \ldots, z_n$. Consider the $n!$ automorphisms of K_1 which are obtained by permuting $z_1, \ldots, z_n$. These automorphisms form a group isomorphic to S_n and the fixed field of this group is $k_1 = F(b_1, \ldots, b_n)$, also, $[K_1:k_1] = n!$ (see Exercise 4). Suppose that, for $i = 1, \ldots, n$, y_i is the same symmetric function of $a_1, \ldots, a_n$ as b_i is of $z_1, \ldots, z_n$. Define the mapping σ from $F[y_1, \ldots, y_n]$ into $F[b_1, \ldots, b_n]$ by $\sigma(g(y_1, \ldots, y_n)) = g(b_1, \ldots, b_n)$: σ is clearly a homomorphism onto $F[b_1, \ldots, b_n]$. Suppose $\sigma(g(y_1, \ldots, y_n)) = 0$. Then $g(b_1, \ldots, b_n) = 0$ and if we replace each b_i by its expression $b_i(z_1, \ldots, z_n)$ in terms of $z_1, \ldots, z_n$ we have $g(b_1(z_1, \ldots, z_n), \ldots, b_n(z_1, \ldots, z_n)) = 0$. Since $z_1, \ldots, z_n$ are independent indeterminates, the result of substituting any elements of any field containing F for $z_1, \ldots, z_n$ will yield 0. Hence $g(b_1(a_1, \ldots, a_n), \ldots, b_n(a_1, \ldots, a_n)) = 0$, that is, $g(y_1, \ldots, y_n) = 0$. Therefore σ is an isomorphism. It can be extended to an isomorphism from k onto k_1. Since K is a splitting field over $k = F(y_1, \ldots, y_n)$ of $f(x)$ and K_1 is a splitting field over $k_1 = F(b_1, \ldots, b_n)$ of $(\sigma f)(x) = x^n + b_1 x^{n-1} + \cdots + b_n$ the isomorphism from k onto k_1 can be extended to one from K onto K_1. Therefore $[K:k] = [K_1:k_1] = n!$. ∥

The fact that $f(x) = 0$ is not solvable by radicals is a consequence of the following result.[3]

THEOREM 19. *If $n \geqslant 5$ then S_n is not solvable.*

10. Infinite Galois extensions

Let k be a field and let K be a normal and separable extension of k. Recall that this means that K/k is algebraic, that every $a \in K$ is a simple root of Irr (k,a), and that every irreducible polynomial in $k[x]$ which has one root in K splits in $K[x]$. We shall not assume that K is a finite extension of k and,

[3] A proof of this theorem may be found in Appendix 1.

in fact, our interest in this section is in the case when K is an infinite extension of k.

PROPOSITION 1. *If σ is a k-isomorphism of K into itself then σ is a k-automorphism of K.*

Proof. Let $a \in K$ and let $f(x) = \text{Irr}(k,a)$. Since $f(x)$ has one root in K, namely a, it splits in $K[x]$. Therefore, there is a splitting field L of $f(x)$ over k such that $k \subseteq L \subseteq K$: L/k is finite and normal and $a \in L$. We know that $k \subseteq \sigma(L) \subseteq L$ and that $[\sigma(L):k] = [L:k]$ and so $\sigma(L) = L$. Hence there is an element $b \in L$ such that $\sigma(b) = a$. ‖

PROPOSITION 2. *Let $k \subseteq L \subseteq K$ and let σ be a k-isomorphism of L into K. Then there is a k-automorphism τ of K such that $\tau(a) = \sigma(a)$ for all $a \in L$.*

Proof. Let $\mathfrak{S}$ be the set of all ordered pairs (F,ρ) where F is a field with $L \subseteq F \subseteq K$ and ρ is a k-isomorphism of F into K such that $\rho(a) = \sigma(a)$ for all $a \in L$. If (F_1,ρ_1) and (F_2,ρ_2) are in $\mathfrak{S}$ we write $(F_1,\rho_1) \leqslant (F_2,\rho_2)$ if $F_1 \subseteq F_2$ and if $\rho_2(a) = \rho_1(a)$ for all $a \in F_1$. This makes $\mathfrak{S}$ into an inductively ordered set which is not empty since (L,σ) belongs to $\mathfrak{S}$. By Zorn's lemma, $\mathfrak{S}$ has a maximal element (F',ρ'). We claim that $F' = K$. For, suppose that there is an $a \in K$ with $a \notin F'$. Then Irr (k,a) splits in $K[x]$ and we have Irr $(k,a) = f(x)g(x)$ where $f(x) = \text{Irr}(F',a)$ and $g(x) \in F'[x]$. Then

$$\text{Irr }(k,a) = (\rho'f)(x)(\rho'g)(x)$$

and so $(\rho'f)(x)$ splits in $K[x]$. Let a_1 be a root of $(\rho'f)(x)$ in K. There is an isomorphism ρ'' of $F'(a)$ onto $(\rho'(F'))(a_1)$ such that $\rho''(a) = a_1$ and $\rho''(b) = \rho'(b)$ for all $b \in F'$. Thus $(F'(a),\rho'') \in \mathfrak{S}$ and is strictly greater than (F', ρ'), which is a contradiction. Thus we do have $F' = K$ and, by Proposition 1, ρ' is a k-automorphism of K, and so is the desired τ. ‖

THEOREM 20. *The field K is a Galois extension of k, that is, k is the fixed field of $G(K/k)$.*

Proof. Let a be an element of the fixed field of $G(K/k)$. As in the proof of Proposition 1 there is a finite normal extension L of k with $k \subseteq L \subseteq K$ and $a \in L$. Let $\sigma \in G(L/k)$. By Proposition 2 there is an element $\tau \in G(K/k)$ such that $\tau(b) = \sigma(b)$ for all $b \in L$. In particular, $\tau(a) = \sigma(a) = a$ and so a is in the fixed field of $G(L/k)$. It follows from Theorem 4 that $a \in k$. ‖

We shall now prove the converse of Theorem 20, that is, *if K is a Galois extension of k then K/k is normal and separable.* Since K/k is Galois it is algebraic. Let $a \in K$ and let $f(x) = \text{Irr}(k,a)$. Let $a = a_1, a_2, \ldots, a_n$ be the distinct images of a under the various k-automorphisms of K. Then $f(a_i) = 0$ for $i = 1, \ldots, n$ and so if $g(x) = (x - a_1) \cdots (x - a_n)$ we see that $g(x)$ divides $f(x)$ in $K[x]$. However, each coefficient of $g(x)$ is left fixed by each element of $G(K/k)$ and so $g(x) \in k[x]$. But the $g(x)$ divides $f(x)$ in $k[x]$ and since $f(x)$ is monic and irreducible in $k[x]$ it follows that $f(x) = g(x)$. Therefore a is separable over k and we conclude that K/k is separable. If we view $f(x)$ as the monic constant multiple of an arbitrary nonconstant irreducible polynomial in $k[x]$ which has one root in K we see that such polynomials split in $K[x]$. Therefore, K/k is normal.

Thus we see that Theorem 4 continues to hold even when k is an infinite extension of k.

Now let K be a Galois extension of k and let H be a subgroup of $G(K/k)$. Let L be the fixed field of H. If the elements σ and τ of $G(K/k)$ belong to the same left coset of H then σ and τ determine the same k-isomorphism of L into K. On the other hand, if σ and τ determine the same k-isomorphism of L into K then $\sigma^{-1}\tau$ leaves each element of L fixed. However, this does not mean that σ and τ determine the same left coset of H, for there may be other elements of $G(K/k)$ besides those in H which leave each element of L fixed. In other words, we may have $H \neq G(K/L)$ (of course, we always have $H \subseteq G(K/L)$). In the proof of Theorem 5 we have shown that this cannot happen when K is a *finite* Galois extension of k, but our proof definitely made use of the fact that $G(K/k)$ is finite. In the next section we shall give an example which shows that we can, indeed, have $H \neq G(K/L)$ when K is an infinite extension of k.

Thus, it turns out that there is not necessarily a one-one correspondence between the fields L such that $k \subseteq L \subseteq K$ and the subgroups of $G(K/k)$ when K is an infinite Galois extension of k. However, as we shall see, the situation can be saved by the introduction of a suitable topology on $G(K/k)$.

We note that if $k \subseteq L \subseteq K$ and if $[L:k] = n$ then $G(K/L)$ has index n in $G(K/k)$. For, if $\sigma_1, \ldots, \sigma_n$ are elements of $G(K/k)$ which, when restricted to L, yield the n distinct k-isomorphisms of L into K, then they represent distinct cosets of $G(K/L)$ in $G(K/k)$. And, if $\sigma \in G(K/k)$ then $\sigma(a) = \sigma_i(a)$ for all $a \in L$ for some i, and so $\sigma G(K/L) = \sigma_i G(K/L)$.

11. The Krull topology

Let K be a Galois extension of k and let $G = G(K/k)$. Let $\mathfrak{N}$ be the family of all subgroups $G(K/F)$ of G where $k \subseteq F \subseteq K$ and $[F:k]$ is finite.

THEOREM 21. *There is a topology on G which is compatible with the group structure of G and which has $\mathfrak{N}$ as a fundamental system of neighborhoods of the identity. With this topology G is a Hausdorff, totally disconnected, compact topological group.*[4]

---

[4] We assume familiarity with these topological concepts. Our general reference for things topological will be [45] and [46] which we refer to as Bourbaki's Chapter I and Bourbaki's Chapter III.

Proof. We see from §1, no. 2 of Bourbaki's Chapter III that we must show that $\mathfrak{N}$ is a filter base and that if $H \in \mathfrak{N}$ and $\sigma \in G$ then $\sigma H \sigma^{-1} \in \mathfrak{N}$. It is certainly true that $\mathfrak{N}$ is not empty and that no element of $\mathfrak{N}$ is the empty set. If, for $i = 1, 2$, $H_i = G(K/F_i) \in \mathfrak{N}$ then

$$H_1 \cap H_2 = G(K/F_i) \cap G(K/F_2) = G(K/F_1F_2) \in \mathfrak{N}$$

and for $\sigma \in G$,

$$\sigma H_1 \sigma^{-1} = \sigma G(K/F_1)\sigma^{-1} = G(K/\sigma(F_1)) \in \mathfrak{N}.$$

To show that G is Hausdorff it suffices to show that

$$\bigcap_{H \in \mathfrak{N}} H = 1.$$

If σ belongs to each H in $\mathfrak{N}$ then σ leaves fixed each element of each subfield of K which is a finite extension of k. Hence $\sigma(a) = a$ for all $a \in K$, that is, $\sigma = 1$.

To show that G is totally disconnected it suffices to show that each $H \in \mathfrak{N}$ is closed as well as open. Let $\sigma \in G$ but $\sigma \notin H$. Then σH is an open set containing σ and $H \cap \sigma H$ is empty. Thus H is closed.

Finally, we shall show that G is compact by showing that every ultra-filter on G converges. Let $\mathfrak{U}$ be an ultrafilter on G. Then if $S_1, \ldots, S_r$ are subsets of G and $S_1 \cup \cdots \cup S_r \in \mathfrak{U}$ it follows that $S_i \in \mathfrak{U}$ for some i.[5] Let $a \in K$ and let $H = G(K/k(a))$. By the remark at the end of the preceding section, H has finite index in G. If $\tau_1 H, \ldots, \tau_s H$ are the distinct left cosets of H in G we have $G = \tau_1 H \cup \cdots \cup \tau_s H \in \mathfrak{U}$ and so $\tau_i H \in \mathfrak{U}$ for some i. If $j \neq i$ then $\tau_j H \cap \tau_i H$ is empty and so $\tau_j H \notin \mathfrak{U}$. We now define $\sigma(a)$ to be $\tau_i(a)$. We shall show that $\sigma(a)$ may actually be defined in a like manner using any finite extension of k in K which contains a. Suppose that $a \in L$ where $k \subseteq L \subseteq K$ and $[L:k]$ is finite. Let $H' = G(K/L)$. By the argument used above there is exactly one left coset $\tau' H'$ of H' in G which is in $\mathfrak{U}$. Note that $H' \subseteq H$. Since $\tau_i H$ and $\tau' H'$ are both in $\mathfrak{U}$, $\tau_i H \cap \tau' H'$ is not empty. If ρ is in this intersection then $\rho^{-1}\tau' \in H' \subseteq H$ and $\tau_i^{-1}\rho \in H$ so that $\tau_i^{-1}\tau' = (\tau_i^{-1}\rho)(\rho^{-1}\tau') \in H$. Thus $\tau' H = \tau_i H$ and $\tau' H' \subseteq \tau_i H$. Therefore $\tau'(a) = \tau_i(a) = \sigma(a)$, which was to be shown. Since two arbitrary elements of K belong to some finite extension of k in K, σ takes sums into sums and products into products. Since $\sigma(1) = 1$, σ is an isomorphism of K into itself and by Proposition 1 of Section 10, $\sigma \in G$. We shall prove that the ultrafilter $\mathfrak{U}$ converges to σ by showing that every neighborhood of σ is in $\mathfrak{U}$. To do this it is sufficient to show that for every $H \in \mathfrak{N}$, we have $\sigma H \in \mathfrak{U}$. Let $H = G(K/F) \in \mathfrak{N}$. Then $[F(a):k]$ is finite and we let $H' = G(K/F(a))$. Then $H' \subseteq H$ and by the way in which σ is determined we have $\sigma H' \in \mathfrak{U}$. But $\sigma H' \subseteq \sigma H$ and so $\sigma H \in \mathfrak{U}$. ∥

[5] See Bourbaki's Chapter I, §5, no. 7.

The topology on $G(K/k)$ which we have described in the above theorem is known as the *Krull topology*. From now on when we refer to the Galois group $G(K/k)$ we shall mean the topological group obtained when $G(K/k)$ is provided with its Krull topology. Note that if K is a finite extension of k then the Krull topology of $G(K/k)$ is the discrete topology and we cannot expect any new information in finite Galois theory to result from these topological considerations.

Now let K be a Galois extension of k and let $G = G(K/k)$. Let H be a subgroup of G and denote the closure of H by $\bar{H}$.

PROPOSITION. *If L is the fixed field of H then $G(K/L) = \bar{H}$.*

Proof. Let $\sigma \in \bar{H}$ and let $a \in L$. Let $H' = G(K/k(a))$. Then $H \cap \sigma H'$ is not empty. Let τ be an element of this intersection. Then τ leaves each element of L fixed and $\sigma^{-1}\tau$ leaves each element of $k(a)$ fixed. Therefore, $\sigma(a) = \tau(a) = a$ so that σ leaves each element of L fixed. Thus $\bar{H} \subseteq G(K/L)$.

Now let $\sigma \in G(K/L)$. To show that $\sigma \in \bar{H}$, we must show that for each $H' \in \mathfrak{N}$ we have $H \cap \sigma H'$ not empty. Since the fixed field of H' is a finite separable extension of k it is of the form $k(a)$ for some $a \in K$. Let F be a finite normal extension of L in K which contains a. If $\rho \in H$ then the restriction of ρ to F is an L-automorphism of F and lies in $G(F/L)$. It follows that all of $G(F/L)$ is obtained in this way since L is exactly the fixed field of H. The restriction of σ to F is some element of $G(F/L)$ and there is an element $\tau \in H$ such that $\sigma(b) = \tau(b)$ for all $b \in F$. In particular, $\sigma(a) = \tau(a)$ and therefore $\sigma^{-1}\tau$ leaves each element of $k(a)$ fixed, that is, $\sigma^{-1}\tau \in H'$. Then $\tau \in H \cap \sigma H'$. ‖

Now let H be a closed subgroup of G and let $L(H)$ be the fixed field of H. By the proposition, if $L(H_1) = L(H_2)$ then $H_1 = \bar{H}_1 = G(K/L(H_1)) = G(K/L(H_2)) = \bar{H}_2 = H_2$. Hence the mapping

$$H \to L(H)$$

from the set of closed subgroups of G into the set of all fields L with $k \subseteq L \subseteq K$ is one-one. We shall now show that it is onto. Let $k \subseteq F \subseteq K$. We shall show that $F = L(G(K/F))$, that is, that F is the fixed field of $G(K/F)$. It then follows from the proposition that $G(K/F)$ is a closed subgroup of G. Since F is certainly contained in the fixed field of $G(K/F)$ we must show that if $a \notin F$ then $\sigma(a) \neq a$ for some $\sigma \in G(K/F)$. Let F' be a finite normal extension of F in K such that $a \in F'$. Since $a \notin F$ there is an F-automorphism τ of F' such that $\tau(a) \neq a$. By Proposition 2 of Section 10 there is an F-automorphism σ of K such that $\sigma(b) = \tau(b)$ for all $b \in F'$. Then $\sigma \in G(K/F)$ and $\sigma(a) \neq a$.

The inverse of the mapping $H \to L(H)$ is $L \to G(K/L)$. We can now state the more general form of the fundamental theorem of Galois theory.

THEOREM 22. *Let K be a Galois extension of k. Then there is a one-one correspondence between the closed subgroups of $G(K/k)$ and the fields L such that $k \subseteq L \subseteq K$. This correspondence is $L \leftrightarrow G(K/L)$.*

Let $k \subseteq L \subseteq K$ and suppose that L/k is finite. Then it is clear that L/k is normal if and only if $G(K/L)$ is a normal subgroup of $G(K/k)$, and if L/k is normal then

$$G(L/k) \cong G(K/k)/G(K/L).$$

This is also true when L/k is not necessarily finite (see Exercise 47).

Remark. In the proof of Theorem 21 it was shown that the only element of $G(K/k)$ which belongs to all of the subgroups in the family $\mathfrak{N}$ is 1. In fact, we can show that

$$\bigcap_{\substack{H \in \mathfrak{N} \\ H \text{ normal}}} H = 1.$$

For, suppose that σ belongs to every normal subgroup of $G(K/k)$ which belongs to $\mathfrak{N}$. If $\sigma \neq 1$ there is an $a \in K$ such that $\sigma(a) \neq a$. Let L be a finite normal extension of k in K which contains a. There is such an extension by Theorem 19, Chapter 1. Then $G(K/L) \in \mathfrak{N}$ and is a normal subgroup of $G(K/k)$ but $\sigma \notin G(K/L)$. Hence we must have $\sigma = 1$.

Now we shall give the example promised in Section 10. This example will illustrate two important facts. First, there may be relatively few subgroups of $G(K/k)$ which are closed and, second, there may be subgroups of finite index in $G(K/k)$ which are not closed.

Let C be an algebraic closure of the field Q of rational numbers and let K be the composite of all of the quadratic extensions of Q in C. Then $K = Q(S)$ where $S = \{\sqrt{p} \mid p = -1 \text{ or } p \text{ is a prime}\}$. If $\sigma \in G(K/Q)$ then it is clear that $\sigma^2 = 1$. Hence $G(K/Q)$ may be considered as a vector space over $GF(2)$. As such, it has a basis B. If T is any subset of S then there is an element $\sigma \in G(K/Q)$ such that $\sigma(\sqrt{p}) = -\sqrt{p}$ for all $\sqrt{p} \in T$ and $\sigma(\sqrt{p}) = \sqrt{p}$ for all $\sqrt{p} \notin T$. Therefore, $\#G(K/Q) = 2^c$ where c is the cardinal number of the set of integers. If B were a denumerable set then, since every element of $G(K/Q)$ is a linear combination of a finite number of elements of B, $G(K/Q)$ would be denumerable. Hence B is not denumerable. If we remove one element from B then the remaining elements generate a subgroup of index two in $G(K/Q)$. Hence $G(K/Q)$ has a nondenumerable number of subgroups of index two. However, it is clear that there is only a denumerable number of quadratic extensions of Q in K. Thus, there is only a denumerable number of closed subgroups of index two in $G(K/Q)$.

12. Inverse limits

A partially ordered set S is called a *directed set* if for each pair of elements $\alpha, \beta \in S$ there is an element $\gamma \in S$ such that $\alpha \leqslant \gamma$ and $\beta \leqslant \gamma$.

Let S be a directed set and let $\{G_\alpha\}$ be a family of topological groups which is indexed by the set S. Suppose that there corresponds to each pair of elements $\alpha, \beta \in S$ with $\alpha \leqslant \beta$ a continuous homomorphism

$$\phi_{\beta\alpha}\colon G_\beta \to G_\alpha$$

such that $\phi_{\alpha\alpha}$ is the identity mapping of G_α for each $\alpha \in S$ and

$$\phi_{\gamma\alpha} = \phi_{\beta\alpha}\phi_{\gamma\beta} \qquad \text{whenever} \qquad \alpha \leqslant \beta \leqslant \gamma.$$

The resulting composite object $(S,\{G_\alpha\},\{\phi_{\alpha\beta}\})$ is called an *inverse system of groups*, and will usually be denoted simply by $\{G_\alpha\}$.

Consider the *direct product* of the groups G_α, that is, the set of all functions $\{\sigma_\alpha\}$ where $\sigma_\alpha \in G_\alpha$ for all $\alpha \in S$ and with the binary operation $\{\sigma_\alpha\}\{\tau_\alpha\} = \{\sigma_\alpha\tau_\alpha\}$. We denote the resulting group by $\prod G_\alpha$. In the direct product of the G_α there are certain elements $\{\sigma_\alpha\}$ such that for all $\beta, \gamma \in S$ with $\beta \leqslant \gamma$ we have $\phi_{\gamma\beta}(\sigma_\gamma) = \sigma_\beta$. An example of such an element is $\{\sigma_\alpha\}$ where $\sigma_\alpha = 1$ for all $\alpha \in S$. Let G be the set of all such elements of $\prod G_\alpha$. It is easily verified that G is a subgroup of $\prod G_\alpha$. If we give to $\prod G_\alpha$ its Cartesian product topology then it becomes a topological group. As a subset of $\prod G_\alpha$, G has an induced topology.

PROPOSITION 1. *The group G, together with the topology induced on it by the Cartesian product topology on $\prod G_\alpha$, is a topological group.*

This is a special case of a general result (Bourbaki's Chapter III, §2, no. 1). The resulting topological group G is called the *inverse limit* of the inverse system $\{G_\alpha\}$ and we write

$$G = \varprojlim G_\alpha.$$

Some authors call the inverse limit the *projective limit*.

For the rest of this section we shall suppose that G is the inverse limit of an inverse system $\{G_\alpha\}$. There is a homomorphism $\phi_\beta\colon G \to G_\beta$ which is given by $\phi_\beta(\{\sigma_\alpha\}) = \sigma_\beta$. We note that for $\beta, \gamma \in S$ with $\beta \leqslant \gamma$ we have $\phi_{\gamma\beta}\phi_\gamma(\{\sigma_\alpha\}) = \phi_{\gamma\beta}(\sigma_\gamma) = \sigma_\beta = \phi_\beta(\{\sigma_\alpha\})$. Since ϕ_β is the restriction to G of the βth projection mapping of $\prod G_\alpha$ it is continuous.

PROPOSITION 2. *If, for each $\alpha \in S$, G_α is a Hausdorff space, then G is a closed subset of $\prod G_\alpha$.*

Proof. Let $\{\sigma_\alpha\}$ be an element of $\prod G_\alpha$ which is not in G. We must find a neighborhood of $\{\sigma_\alpha\}$ in $\prod G_\alpha$ which does not intersect G. Since $\{\sigma_\alpha\} \notin G$ there must be a pair of elements $\beta, \gamma \in S$ with $\beta \leqslant \gamma$ such that $\phi_{\gamma\beta}(\sigma_\gamma) \neq \sigma_\beta$. Since G_β is a Hausdorff space there is an open neighborhood M_β of σ_β and an open neighborhood N_β of $\phi_{\gamma\beta}(\sigma_\gamma)$ such that $M_\beta \cap N_\beta$ is empty. Let $N_\gamma = \phi_{\gamma\beta}^{-1}(N_\beta)$. Since $\phi_{\gamma\beta}$ is continuous, N_γ is open in G_γ and $\sigma_\gamma \in N_\gamma$. The subset $\prod X_\alpha$ of $\prod G_\alpha$, where $X_\beta = M_\beta$, $X_\gamma = N_\gamma$, and $X_\alpha = G_\alpha$ when

$\alpha \neq \beta, \gamma$, is an open set in $\prod G_\alpha$ and it contains $\{\sigma_\alpha\}$. Furthermore, if $\{\tau_\alpha\} \in \prod X_\alpha$ then $\tau_\gamma \in N_\gamma$, $\tau_\beta \in M_\beta$, and $\phi_{\gamma\beta}(\tau_\gamma) \in N_\beta$, and so $\phi_{\gamma\beta}(\tau_\gamma) \neq \tau_\beta$. Hence $\prod X_\alpha$ does not intersect G. Therefore G is closed. $\parallel$

PROPOSITION 3. *If, for each $\alpha \in S$, G_α is a compact Hausdorff space, then G is a compact Hausdorff space.*

Proof. The Cartesian product of compact Hausdorff spaces is compact and Hausdorff. Furthermore, a closed subset of a compact Hausdorff space is compact and Hausdorff. The proposition follows from these general facts and from Proposition 2. $\parallel$

THEOREM 23. *Let G be a compact Hausdorff topological group. Let $\{G_\alpha\}$ be an inverse system of compact Hausdorff groups. Suppose that for each $\beta \in S$ there is a continuous homomorphism ψ_β from G onto G_β such that for all $\beta, \gamma \in S$ with $\beta \leqslant \gamma$ we have $\phi_{\gamma\beta}\psi_\gamma = \psi_\beta$. Suppose further that for all $\sigma, \tau \in G$ with $\sigma \neq \tau$ there is a $\beta \in S$ such that $\psi_\beta(\sigma) \neq \psi_\beta(\tau)$. Then*

$$G \cong \lim_{\leftarrow} G_\alpha,$$

where the isomorphism is topological.

Proof. Let $G' = \lim_{\leftarrow} G_\alpha$. If $\sigma \in G$ we set $\psi(\sigma) = \{\psi_\alpha(\sigma)\}$. For all $\beta, \gamma \in S$ with $\beta \leqslant \gamma$ we have $\phi_{\gamma\beta}\psi_\gamma(\sigma) = \psi_\beta(\sigma)$ and so $\psi(\sigma) \in G'$ for all $\sigma \in G$. Clearly, ψ is a homomorphism from G into G' and is continuous since each ψ_β is continuous. If $\sigma, \tau \in G$ and $\sigma \neq \tau$ then there is a $\beta \in S$ such that $\psi_\beta(\sigma) \neq \psi_\beta(\tau)$ and so $\psi(\sigma) \neq \psi(\tau)$. Thus ψ is one-one.

It remains to show that ψ is *onto*. This will complete the proof since a one-one continuous mapping from one compact Hausdorff space onto another is a homeomorphism. Let $\{\sigma_\alpha\} \in G'$: we must find a $\sigma \in G$ such that $\psi_\beta(\sigma) = \sigma_\beta$ for all $\beta \in S$. If for each $\beta \in S$ we set $\Sigma_\beta = \psi_\beta^{-1}(\sigma_\beta)$ this is the same as showing that

$$\bigcap_{\beta \in S} \Sigma_\beta \quad \text{is not empty.}$$

Now for each $\beta \in S$, Σ_β is a closed subset of the compact Hausdorff space G, and so to show that this intersection is not empty it is sufficient to show that for $\beta_1, \ldots, \beta_n \in S$,

$$\bigcap_{i=1}^n \Sigma_{\beta_i} \quad \text{is not empty.}$$

Since S is a directed set there is a $\gamma \in S$ such that $\beta_i \leqslant \gamma$ for $i = 1, \ldots, n$. Since ψ_γ is *onto* there is a $\sigma \in G$ such that $\psi_\gamma(\sigma) = \sigma_\gamma$. Then, for $i = 1, \ldots, n$, $\sigma_{\beta_i} = \phi_{\gamma\beta_i}(\sigma_\gamma) = \phi_{\gamma\beta_i}\psi_\gamma(\sigma) = \psi_{\beta_i}(\sigma)$ and so $\sigma \in \Sigma_{\beta_i}$. $\parallel$

COROLLARY. *Let G be a compact Hausdorff group and let $\{H_\alpha\}$ be a family of closed normal subgroups of G indexed by a set S. Suppose that for $\alpha, \beta \in S$*

there is a $\gamma \in S$ such that $H_\gamma \subseteq H_\alpha \cap H_\beta$. *Suppose further that* $\cap H_\alpha = 1$. *For each* $\alpha \in S$ *set* $G_\alpha = G/H_\alpha$. *If* $H_\beta \subseteq H_\alpha$ *let* $\phi_{\beta\alpha}$ *be the homomorphism from* G_β *onto* G_α *induced by the natural homomorphism from* G *onto* G/H_α. *Then* S *can be partially ordered in such a way that* $(S,\{G_\alpha\},\{\phi_{\beta\alpha}\})$ *is an inverse system of groups and*

$$G \cong \varprojlim G_\alpha.$$

Proof. Define $\alpha \leqslant \beta$ if and only if $H_\beta \subseteq H_\alpha$. This makes S into a directed set. For each $\alpha \in S$, G_α is a compact Hausdorff group and $(S,\{G_\alpha\},\{\phi_{\beta\alpha}\})$ is an inverse system of groups. For each $\alpha \in S$ let ψ_α be the natural homomorphism from G onto $G/H_\alpha = G_\alpha$. Then for $\beta \leqslant \gamma$, $\phi_{\gamma\beta}\psi_\gamma = \psi_\beta$. Let $\sigma, \tau \in G$ with $\sigma \neq \tau$. Then $\sigma^{-1}\tau$ does not belong to $\cap H_\alpha = 1$ and so there is a $\beta \in S$ such that $\sigma^{-1}\tau \notin H_\beta$. Then $\sigma H_\beta \neq \tau H_\beta$, that is, $\psi_\beta(\sigma) \neq \psi_\beta(\tau)$. The corollary now follows from the theorem. ‖

Let K be a Galois extension of k and let $\mathfrak{F}$ be the family of all finite normal extensions L of k with $k \subseteq L \subseteq K$. For each $L \in \mathfrak{F}$ we know from the preceding section that $G(K/L)$ is a closed normal subgroup of the compact Hausdorff group $G(K/k)$. Furthermore, we know that if $L, F \in \mathfrak{F}$ then $G(K/L) \cap G(K/F)$ is the Galois group of K over some field in $\mathfrak{F}$. By the remark at the end of the last section we have

$$\bigcap_{L \in \mathfrak{F}} G(K/L) = 1.$$

Thus we are in a position to apply the corollary to Theorem 23. For each $L \in \mathfrak{F}$ we have $G(K/k)/G(K/L) \cong G(L/k)$ and by Exercise 47, the topological quotient group $G(L/k)$ has its Krull topology. Therefore, as a finite Hausdorff topological group, $G(L/k)$ is discrete. (Actually, this follows from a general proposition since $G(K/L)$ is open as well as closed in $G(K/k)$: see Bourbaki's Chapter III, Proposition 10.) We consider $\mathfrak{F}$ as a partially ordered set, ordered by set inclusion. If $L \subseteq F$, let $\phi_{F/L}$ be the homomorphism from $G(F/k)$ onto $G(L/k)$ induced by the natural homomorphism from $G(K/k)$ onto $G(K/k)/G(K/L)$. This is precisely the natural homomorphism from $G(F/k)$ onto $G(L/k) \cong G(F/k)/G(F/L)$. Then $(\mathfrak{F},\{G(L/k)\},\{\phi_{F/L}\})$ is an inverse system of groups and we have

THEOREM 24. *If* K *is a Galois extension of* k *then* $G(K/k)$ *is the inverse limit of the inverse system* $(\mathfrak{F},\{G(L/k)\},\{\phi_{F/L}\})$:

$$G(K/k) \cong \varprojlim G(L/k).$$

In Appendix 2 we shall give an example in which we use these ideas to give an explicit description of the Galois group of a certain infinite Galois extension.

EXERCISES

Section 1

1. Let L/k be separable and let K be the smallest normal extension of k such that $L \subseteq K$. Show that K/k is Galois.

2. Consider the field $k(x)$ of rational functions over a field k. Show that the mappings of $k(x)$ into itself given by

$$f(x) \to f(x), \qquad f(x) \to f\left(\frac{1}{1-x}\right), \qquad f(x) \to f\left(\frac{x-1}{x}\right)$$

form a group of automorphisms of $k(x)$. Determine the fixed field of this group and the degree of $k(x)$ over this fixed field.

3. Again consider $k(x)$, but assume that char $k = 0$. Show that the mapping $f(x) \to f(x+1)$ is an automorphism of $k(x)$ and generates an infinite cyclic group G of automorphisms of $k(x)$. Determine the fixed field F of G. What is $[k(x):F]$?

4. Let $K = k(x_1, \ldots, x_n)$ be the field of rational functions in n independent indeterminates over a field k. Let $a_1, \ldots, a_n$ be the elementary symmetric functions in $x_1, \ldots, x_n$. Let G be the group of permutations of $x_1, \ldots, x_n$ and consider each element of G as an automorphism of K in the natural way. Let $F = k(a_1, \ldots, a_n)$.

(a) Show that F is contained in the fixed field of G and so $[K:F] \geqslant n!$.

(b) Show that $[K:F] \leqslant n!$, and conclude that F is the fixed field of G and that $[K:F] = n!$. (Hint. Define $F = F_n \subset F_{n-1} \subset F_{n-2} \subset \cdots \subset F_2 \subset F_1 = K$ by $F_j = F(x_{j+1}, \ldots, x_n)$ for $j = 1, \ldots, n-1$, and show that $[F_{j-1}:F_j] \leqslant j$ by showing that x_j is a root of a polynomial in $F_j[x]$ of degree $\leqslant j$.)

Section 2

5. Let K be a finite Galois extension of k and let $k \subseteq L \subseteq K$ and $k \subseteq F \subseteq K$. Show that $G(K/LF) = G(K/L) \cap G(K/F)$ and $G(K/L \cap F) = G(K/L) \cup G(K/F)$ (this is the smallest subgroup of $G(K/k)$ which contains both $G(K/L)$ and $G(K/F)$). What can we conclude, then, from the fact that $G(K/L) \cap G(K/F) = 1$?

6. Let K be a finite Galois extension of k and let $k \subseteq F \subseteq K$. Let L be the smallest field such that $F \subseteq L \subseteq K$ and L/k is normal. Show that

$$G(K/L) = \bigcap_{\sigma \in G(K/k)} \sigma G(K/F) \sigma^{-1}.$$

Give a proof of the second to the last statement in Theorem 5 based on this fact.

7. Let K be an extension of $GF(p^n)$ with $[K:GF(p^n)] = m$. Show that K is a Galois extension of $GF(p^n)$. Let the automorphism σ of K be given by $\sigma(a) = a^{p^n}$ for all $a \in K$. Show that $G(K/GF(p^n))$ is cyclic and generated by σ.

8. Let k be a finite field and K a finite extension of k. Show that every nonzero element of k is the norm of exactly $(K^*:k^*)$ elements of K^*. Show by example that this is not true, in general, when k is an infinite field. (Hint. Make use of the generator of $G(K/k)$ given in Exercise 7.)

9. A field k is called a *quasi-finite field* if k is perfect and if it has, in any algebraic closure, exactly one extension of degree n for each integer $n > 0$. Let k be a quasi-finite field and let K be a finite extension of k. Show that K is a cyclic extension of k.

10. Let K be a finite Galois extension of k and let p be a prime. Assume that $p^r \mid [K:k]$ but $p^{r+1} \nmid [K:k]$. Show that there are subfields L_i of K, $i = 0, 1, \ldots, r$, such that $k \subseteq L_r \subset L_{r-1} \subset \cdots \subset L_1 \subset L_0 = K$, L_i/L_{i+1} is normal, $[L_i : L_{i+1}] = p$, and $p \nmid [L_r : k]$.

11. Let K be a finite Galois extension of k and let $G = G(K/k)$. If $a \in K$ has the property that $\{\sigma(a) \mid \sigma \in G\}$ is a basis of K/k, then this basis is called a *normal basis* of K/k. Let $G = \{\sigma_1, \ldots, \sigma_n\}$. Show that $\{\sigma(a) \mid \sigma \in G\}$ is a normal basis of K/k if and only if the matrix $[\sigma_i \sigma_j(a)]$ has nonzero determinant.

12. Show that if K/k is cyclic then K/k has a normal basis.

13. Show that if k is an infinite field and if K/k is finite Galois, then K/k has a normal basis. Combining this with Exercises 7 and 12 we see that, in all cases, K/k has a normal basis.

14. Let K be a finite Galois extension of k. Show that K/k has a normal basis $\{\sigma(a) \mid \sigma \in G\}$ with $T_{K/k}(a) = 1$.

15. Let K be a finite separable extension of k and let F be the smallest normal extension of k which contains K. Let $a \in F$ be such that $\{\sigma(a) \mid \sigma \in G(F/k)\}$ is a normal basis of F/k and let $b = T_{F/K}(a)$. Show that $K = k(b)$.

16. Let K and F be finite Galois extensions of k with $K \cap F = k$ and let $a \in k$ and $b \in F$ be such that $\{\sigma(a) \mid \sigma \in G(K/k)\}$ and $\{\sigma(b) \mid \sigma \in G(F/K)\}$ are normal bases of K/k and F/k, respectively. Show that $\{\sigma(ab) \mid \sigma \in G(KF/k)\}$ is a normal basis of KF/k.

Section 3

17. Determine the subfields of $Q(\sqrt[4]{2}, i)$. Which of these subfields are Galois extensions of Q? Show that this field is a splitting field of $x^4 - 2$ over Q.

18. Determine the Galois group and all of the subfields of the splitting field of $(x^3 - 2)(x^2 - 3)$ over the field of rational numbers.

Section 4

19. The results of this exercise are needed in the proof of Proposition 4 of Section 4. A polynomial $f(x) \in J[x]$ is called *primitive* if the g.c.d. of its coefficients is one.

(a) (Gauss' Lemma.) Show that the product of primitive polynomials in $J[x]$ is primitive.

(b) If $f(x) \in Q[x]$ then we can write $f(x) = c(f)f_1(x)$ where $f_1(x)$ is a primitive polynomial in $J[x]$. Show that if $f(x) = g(x)h(x)$, where $g(x), h(x) \in Q[x]$, then $c(f) = c(g)c(h)$.

(c) Let $f(x) \in J[x]$ and suppose that $f(x)$ factors in $Q[x]$. Show that $f(x)$ factors in $J[x]$ into factors of the same degrees.

(d) Let $f(x)$ be a monic polynomial in $J[x]$. Then there are polynomials $p_1(x), \ldots, p_n(x)$ such that for each i, $p_i(x)$ is monic, has integer coefficients, is irreducible in $Q[x]$, and $f(x) = p_1(x) \cdots p_n(x)$.

20. Show that for each positive integer n,

$$F_n(x) = \prod_{d \mid n} (x^d - 1)^{\mu(n/d)}.$$

(Hint. Use the Möbius inversion formula.)

21. Show that

$$F_6(x) = x^2 - x + 1, \qquad F_8(x) = x^4 + 1,$$
$$F_9(x) = x^6 + x^3 + 1, \qquad F_{10}(x) = x^4 - x^3 + x^2 - x + 1,$$
$$F_{12}(x) = x^4 - x^2 + 1,$$

and in general, for a prime p, that

$$F_p(x) = x^{p-1} + x^{p-2} + \cdots + x + 1.$$

22. Show that for $n > 1$,

$$F_n(1) = \begin{cases} p & \text{if } n \text{ is a power of a prime } p \\ 1 & \text{otherwise.} \end{cases}$$

23. Determine the Galois group of K_5 and the fields between Q and K_5. Express these fields in terms of square roots. Do the same thing for K_8.

24. Do the same as in Exercise 23 for K_7. What is Irr $(Q, \zeta + \zeta^{-1})$ where ζ is a primitive 7th root of unity?

25. Let $n > 2$ and let ζ be a primitive nth root of unity. Show that $[Q(\zeta + \zeta^{-1}): Q] = \phi(n)/2$.

26. Let $n > 2$ and $(h,n) = 1$. Show that $[Q(\cos 2\pi h/n): Q] = \phi(n)/2$. If $n \neq 4$ show that

$$[Q(\sin 2\pi h/n): Q] = \begin{cases} \phi(n) & \text{if } (n,8) < 4 \\ \phi(n)/4 & \text{if } (n,8) = 4 \\ \phi(n)/2 & \text{is } (n,8) > 4. \end{cases}$$

27. Let $n > 4$ and $(h,n) = 1$. Show that

$$[Q(\tan 2\pi h/n): Q] = \begin{cases} \phi(n) & \text{if } (n,8) < 4 \\ \phi(n)/2 & \text{if } (n,8) = 4 \\ \phi(n)/4 & \text{if } (n,8) > 4. \end{cases}$$

28. Find a basis of K_n over Q and determine the discriminant of this basis. Find a normal basis of K_n/Q.

29. Let k be an arbitrary field and let ζ be a primitive nth root of unity in an algebraic closure of k (see the beginning of Section 6). Show that $k(\zeta)$ is an Abelian extension of k: in fact, determine $G(k(\zeta)/k)$.

Section 5

30. Let $f \in C^n(G,A)$. Show that $\delta\delta f = 0$ in the cases $n = 0, 1$, and 2.

31. Let K be a cyclic extension of k and let σ be a generator of $G(K/k)$. Let F be an extension of k with $k \subseteq F \subseteq K$ and $[K:F] = m$. Let $a \in k$, $a \neq 0$, and assume that $a^m = N_{K/k}(b)$ for some $b \in K$. Show that there is an element $c \in F$ such that $N_{F/k}(c) = a$.

32. Let K and F be cyclic extensions of k with $[K:k] = [F:k] = p$. Let $k \subseteq L \subseteq KF$ with $[L:k] = p$. Let $a \in k$ and suppose that $a = N_{K/k}(b) = N_{F/k}(c)$ where $b \in K$ and $c \in F$. Show that there is an element $d \in L$ such that $a = N_{L/k}(d)$.

33. Let G be a group and A a G-module. Write A multiplicatively. Let $f \in Z^2(G,A)$. Let G' be the set of all ordered pairs (σ,a) where $\sigma \in G$ and $a \in A$. Define

$$(\sigma,a)(\tau,b) = (\sigma\tau, f(\sigma,\tau)\tau(a)b).$$

Show that G' is a group with respect to this operation. Show that $\{(1,a) \mid a \in A\}$ is a normal subgroup A' of G' with $A \cong A'$ and that $G'/A' \cong G$. The group G' is called a *group extension* of A by G.

34. Let $h \in Z^2(G,A)$ and let H' be the group extension of A by G determined by h. We may consider A' as a subgroup of H'. Show that G' and H' are isomorphic by an isomorphism that maps each element of A' onto itself if and only if f and h determine the same element of $H^2(G,A)$.

35. Let G be a group and A a G-module. If $f \in Z^2(G,A)$ show that there is an element $g \in Z^2(G,A)$ such that f and g determine the same element of $H^2(G,A)$ and $g(\sigma,\tau) = 1$ whenever either $\sigma = 1$ or $\tau = 1$.

36. Let G be a group and A a G-module. Let H be a subgroup of G and define $\rho: C^2(G,A) \to C^2(H,A)$ by $(\rho f)(\sigma,\tau) = f(\sigma,\tau)$ for all $\sigma, \tau \in H$. Show that ρ is a homomorphism which maps $Z^2(G,A)$ into $Z^2(H,A)$ and $B^2(G,A)$ into $B^2(H,A)$ and so induces a homomorphism $\rho^*: H^2(G,A) \to H^2(H,A)$. Now assume that H is a normal subgroup of G and let $A' = \{a \in A \mid \sigma(a) = a$ for all $\sigma \in H\}$. Show that A' is, in a natural way, a (G/H)-module. Define $\lambda: C^2(G/H,A') \to C^2(G,A)$ by $(\lambda f)(\sigma,\tau) = f(\sigma H,\tau H)$. Show that λ is a homomorphism which maps $Z^2(G/H,A')$ into $Z^2(G,A)$ and $B^2(G/H,A')$ into $B^2(G,A)$ and so induces a homomorphism $\lambda^*: H^2(G/H,A') \to H^2(G,A)$.

37. Let K be a finite Galois extension of k, $G = G(K/k)$, F a normal

extension of k with $k \subseteq F \subseteq K$, and $H = G(K/F)$. Show that

$$1 \longrightarrow H^2(G/H, F^*) \xrightarrow{\lambda^*} H^2(G, K^*) \xrightarrow{\rho^*} H^2(H, K^*)$$

is an exact sequence.

Section 6

38. Let p be a prime. Let C be an algebraic closure of a field k and let P be the set (multiplicative group!) of pth roots of unity in C. Show that P has either 1 or p elements. Let $K = k(P)$. Let $x^p - a \in k[x]$, $a \neq 0$, and let L be the splitting field of this polynomial over k in C. Show that either $L = K$ or $K \subset L$ in which case L/K is a cyclic extension of degree p.

39. Let K be a cyclic extension of k with $[K:k] = p^n$ where p is a prime. Let $K = L(a)$ where L/k is cyclic of degree p^{n-1}. Show that $K = k(a)$.

40. Let k be as in Theorem 10 and suppose that K/k is cyclic of degree n. Show that if $K = k(b_1) = k(b_2)$, where b_i is a root of the irreducible polynomial $x^n - a_i \in k[x]$, then $b_2 = b_1{}^t c$ where $c \in k$ and $(t, n) = 1$.

41. Let k be as in Theorem 11 and suppose that K/k is cyclic of degree p. Show that if $K = k(b_1) = k(b_2)$, where b_i is a root of the irreducible polynomial $x^p - x - a_i \in k[x]$, then $b_2 = nb_1 + c$ where $0 < n \leqslant p - 1$ and $c \in k$.

Section 7

42. Let k satisfy the hypothesis of Theorem 14, and let H_1 and H_2 be subgroups of k^* which contain k^{*n} and such that $(H_1 : k^{*n})$ and $(H_2 : k^{*n})$ are finite. Show that the same hold for $H_1 H_2$ and $H_1 \cap H_2$ and that $K_{H_1 H_2} = K_{H_1} K_{H_2}$ and $K_{H_1 \cap H_2} = K_{H_1} \cap K_{H_2}$.

43. Let K be a cyclic Kummer extension of the field k of Theorem 14. Show that if $K = K_H$ and if ak^{*n} is a generator of H/k^{*n} then $K = k(\sqrt[n]{a})$.

Section 8

44. Fill in the details of the proof of Theorem 15.

45. Let k be a quasi-finite field of characteristic p. Show that $(k^+ : \mathscr{P}k^+) = p$.

Sections 10, 11, and 12

46. Let K be a Galois extension of k and let $k \subseteq L \subseteq K$. Show that the topology induced on $G(K/L)$ by the Krull topology of $G(K/k)$ is the Krull topology of $G(K/L)$.

47. Let K be a Galois extension of k. If $k \subseteq L \subseteq K$ show that L/k is normal if and only if $G(K/L)$ is a normal subgroup of $G(K/k)$. Show also

that if L/k is normal then $G(K/k)/G(K/L) \cong G(L/k)$, where the isomorphism is both algebraic and topological.

48. Let K be a Galois extension of k and let G' be the commutator subgroup of $G(K/k)$. Let A be the fixed field of the closure of G' in $G(K/k)$. Show that A is the maximal Abelian extension of k contained in K.

49. Let K be a Galois extension of k and let H be a subgroup of $G(K/k)$. Show that H is dense in $G(K/k)$ if and only if for every finite normal extension L of k with $k \subseteq L \subseteq K$, every k-automorphism of L is obtained by restricting some element of H to L.

50. In the notation of Exercise 49, show that H is dense in $G(K/k)$ if and only if for each finite normal extension L of k with $k \subseteq L \subseteq K$ we have $G(L/k) \cong H/H \cap G(K/L)$.

CHAPTER 3

Introduction to Valuation Theory

1. Definition of valuation: examples

Let k be a field.

DEFINITION. A *valuation* on k is a real-valued function $a \rightarrow |a|$ defined on k which satisfies the following conditions:

(i) $|a| \geqslant 0$ for all $a \in k$ and $|a| = 0$ if and only if $a = 0$,

(ii) $|ab| = |a|\,|b|$ for all $a, b \in k$,

(iii) $|a + b| \leqslant |a| + |b|$ for all $a, b \in k$.

If k is a subfield of the field of complex numbers then ordinary absolute value is a valuation on k. If k is an arbitary field then there is always at least one valuation on k, namely, that given by setting $|a| = 1$ if $a \in k$ and $a \neq 0$ and $|0| = 0$. This valuation is called the *trivial valuation* on k.

We shall now give an example of a valuation which is quite different from the obvious ones given above. Consider the field Q of rational numbers and let p be a prime. Each nonzero element $a \in Q$ can be written uniquely in the form $a = p^n(u/v)$, where u and v are relatively prime integers neither of which is divisible by p, and n is an integer which may be positive, negative, or zero. We now set

$$|a|_p = p^{-n}, \qquad |0|_p = 0.$$

We shall show that $|\ |_p$ is a valuation on Q. Clearly $|a|_p \geqslant 0$ and $|a|_p = 0$ if and only if $a = 0$. Furthermore, if either $a = 0$ or $b = 0$ then $|ab|_p = |a|_p|b|_p$ and $|a + b|_p = |a|_p + |b|_p$. Assume then that $a \neq 0$ and $b \neq 0$ and write $a = p^m(r/s)$ and $b = p^n(u/v)$ where $(rsuv,p) = (r,s) = (u,v) = 1$. Then $ab = p^{m+n}(w/z)$ where $w/z = ru/sv$ and $(wz,p) = 1$. Hence

$$|ab|_p = p^{-(m+n)} = p^{-m}p^{-n} = |a|_p|b|_p.$$

We may assume without loss of generality that $m \leqslant n$. Then

$$a + b = p^m \frac{r}{s} + p^n \frac{u}{v} = p^m \left(\frac{rv + p^{n-m}su}{sv} \right).$$

The denominator sv is not divisible by p while the numerator $rv + p^{n-m}su$ may or may not be divisible by p. However, in any case,

$$|a + b|_p \leqslant p^{-m} = |a|_p = \max (|a|_p, |b|_p)$$
$$\leqslant |a|_p + |b|_p.$$

Thus $|\ |_p$ is a valuation on Q: it is called the *p-adic valuation* on Q. We note that $|\ |_p$ actually satisfies a requirement which is stronger than that of (iii) in the definition of valuation, namely, $|a + b|_p \leqslant \max (|a|_p, |b|_p)$ for all $a, b \in Q$.

DEFINITION.   A valuation $|\ |$ on k is called *non-Archimedean* if

$$|a + b| \leqslant \max (|a|, |b|) \qquad \text{for all } a, b \in k.$$

Any other valuation on k is called *Archimedean*.

As we shall see later on, the fact that a valuation is non-Archimedean has very important algebraic consequences.

DEFINITION.   Two valuations $|\ |_1$ and $|\ |_2$ on k are said to be *equivalent* if either both of them are trivial or $|\ |_1$ is nontrivial and $|a|_1 < 1$ implies $|a|_2 < 1$.

Although it is not immediately apparent that the relation of equivalence of valuations is symmetric we shall show that it is actually an equivalence relation on the set of all valuations on k. It is obvious that the relation is reflexive and transitive: we shall show that it is, indeed, symmetric. If both $|\ |_1$ and $|\ |_2$ are trivial there is nothing to prove. Suppose that $|\ |_1$ is nontrivial and that $|a|_1 < 1$ implies $|a|_2 < 1$. Since $|\ |_1$ is nontrivial there is an element $b \in k$ such that $|b|_1 \neq 0$ or 1. Then either $|b|_1 < 1$ or $|b^{-1}|_1 < 1$. Thus either $|b|_2 < 1$ or $|b^{-1}|_2 < 1$ and $|\ |_2$ is nontrivial. Now suppose that $|a|_2 < 1$. If $|a|_1 > 1$ then $|a^{-1}|_1 < 1$ and so $|a^{-1}|_2 < 1$, or $|a|_2 > 1$. Since this is not true, we must have $|a|_1 \leqslant 1$. Suppose that $|a|_1 = 1$. Since $|\ |_1$ is nontrivial we have $0 < |b|_1 < 1 = |a^n|_1$ for some $b \in k$ and all positive integers n. Hence $|ba^{-n}|_1 < 1$ and so $|ba^{-n}|_2 < 1$, or $|b|_2^{1/n} < |a|_2$, for all positive integers n, and $0 < |b|_2 < 1$. Therefore, $1 = \lim |b|_2^{1/n} \leqslant |a|_2$ which contradicts the fact that $|a|_2 < 1$. Thus $|a|_2 < 1$ implies that $|a|_1 < 1$.

THEOREM 1.   *Let $|\ |_1$ and $|\ |_2$ be equivalent valuations on a field k. Then there is a positive real number c such that $|a|_1 = |a|_2{}^c$ for all $a \in k$.*

Proof.   If $|\ |_1$ and $|\ |_2$ are both trivial there is nothing to prove. Suppose that $|\ |_1$, and therefore $|\ |_2$, is nontrivial. Let $b \in k$ be such that $0 < |b|_1 < 1$.

Then $0 < |b|_2 < 1$ so that there is a positive real number c such that $|b|_1 = |b|_2{}^c$. Now let a be an arbitrary nonzero element of k and let r be the real number such that $|a|_1 = |b|_1{}^r$. Let m/n run through a sequence of rational numbers, each less than or equal to r, which has r as a limit. Then $|a|_1 \leqslant |b|_1^{m/n}$ or $|a^n b^{-m}|_1 \leqslant 1$. Hence $|a^n b^{-m}|_2 \leqslant 1$ or $|a|_2 \leqslant |b|_2^{m/n}$ and so $|a|_2 \leqslant |b|_2{}^r$. If we make similar use of a sequence of rational numbers, each greater than or equal to r, having r as a limit, we will arrive at the conclusion that $|a|_2 \geqslant |b|_2{}^r$. Therefore $|a|_1 = |b|_1{}^r = |b|_2{}^{rc} = |a|_2{}^c$. ∥

2. Valuations on the fields Q and $k(x)$

We shall now set about to determine all the valuations on the field Q of rational numbers and all of the valuations on a field of rational functions which are trivial on k.

LEMMA. *Let M be the subring of a field k which is generated by the multiplicative identity of k. Let $|\ |$ be a valuation on k. Then $|\ |$ is non-Archimedean if and only if it is bounded on M.*

Proof. If $|\ |$ is non-Archimedean then $|1 + \cdots + 1| \leqslant \max(|1|, \ldots, |1|) = 1$ and so $|\ |$ is bounded on M. On the other hand, suppose that there is a real number δ such that $|m \cdot 1| \leqslant \delta$ for all integers m. Then we have, in particular, $\left| \binom{m}{r} a \right| \leqslant \delta|a|$ for all binomial coefficients $\binom{m}{r}$ and all $a \in k$. Hence if $a, b \in k$ and if n is a positive integer we have

$$|a + b|^n = |(a + b)^n| = \left| \sum_{r=0}^{n} \binom{n}{r} a^r b^{n-r} \right|,$$

$$\leqslant \sum_{r=0}^{n} \left| \binom{n}{r} a^r b^{n-r} \right|,$$

$$\leqslant \delta \sum_{r=0}^{n} |a|^r |b|^{n-r}$$

$$\leqslant \delta(n + 1) |a|^n,$$

if $|a| = \max(|a|, |b|)$. Hence

$$|a + b| \leqslant \delta^{1/n}(n + 1)^{1/n} \max(|a|, |b|),$$

and if we let $n \to \infty$ we get $|a + b| \leqslant \max(|a|, |b|)$. ∥

COROLLARY. *If k is a field of characteristic p then every valuation on k is non-Archimedean.*

THEOREM 2. *Let $|\ |$ be a nontrivial valuation on Q. Then $|\ |$ is equivalent to either ordinary absolute value or the p-adic valuation on Q for some prime p.*

Proof. As usual, we denote the ring of integers by J. If we know $|a|$ for all $a \in J$ then $| \ |$ is completely determined as a function on Q, for $|a/b| = |a| \, |b|^{-1}$. In order to prove the theorem we consider two cases.

Case 1. There is an $n \in J$ such that $n > 1$ but $|n| \leqslant 1$. Let $m \in J$ be positive and write

$$m = a_0 + a_1 n + \cdots + a_r n^r, \qquad 0 \leqslant a_i \leqslant n - 1, \qquad a_r \neq 0.$$

For each i, $|a_i| = |1 + \cdots + 1| \leqslant |1| + \cdots + |1| = a_i < n$, so

$$|m| \leqslant \sum_{i=0}^{r} |a_i n^i| = \sum_{i=0}^{r} |a_i| \, |n|^i$$

$$< n \sum_{i=0}^{r} |n|^i \leqslant (r + 1)n.$$

We have $r \leqslant (\log m)/(\log n)$ and so

$$|m| < \left(1 + \frac{\log m}{\log n}\right)n.$$

Now let t be a positive integer and replace m by m^t. This gives

$$|m|^t < \left(1 + t\frac{\log m}{\log n}\right)n,$$

or

$$|m| < \left(1 + t\frac{\log m}{\log n}\right)^{1/t} n^{1/t}$$

for all positive integers t. If we let $t \to \infty$ we obtain $|m| \leqslant 1$ for all positive $m \in J$, and so for all $m \in J$. It then follows from the lemma that $| \ |$ is non-Archimedean. Let

$$I = \{m \in J \mid |m| < 1\}.$$

It is easily seen that I is a prime ideal, say (p), of J, where p is a prime. We have $|p| < 1$ and so there is a positive real number c such that $|p| = p^{-c}$. Also $|m| = 1$ for $m \in J$ if and only if p does not divide m. Now let $a \in Q$ and write $a = p^n(u/v)$ where $u, v \in J$ and $(u,v) = (u,p) = (v,p) = 1$. Then

$$|a| = |p|^n |u| \, |v|^{-1} = |p|^n = |a|_p^c,$$

and so $| \ |$ is equivalent to $| \ |_p$.

Case 2. $|n| > 1$ for all $n > 1$. Let $m, n \in J$, $m > 1$ and $n > 1$, and write

$$m^t = a_0 + a_1 n + \cdots + a_r n^r, \qquad 0 \leqslant a_i \leqslant n - 1, \qquad a_r \neq 0,$$

$$r \leqslant \frac{\log m^t}{\log n},$$

where t is a positive integer. Then

$$|m|^t \leqslant \sum_{i=0}^{r} |a_i| \, |n^i| < n(r+1) \, |n|^r$$

$$\leqslant n \left(1 + \frac{\log m^t}{\log n}\right) |n|^{(\log m^t)/(\log n)}.$$

and therefore

$$|m| \leqslant n^{1/t} \left(1 + t\frac{\log m}{\log n}\right)^{1/t} |n|^{(\log m)/(\log n)}$$

for all positive integers t. If we let $t \to \infty$ we obtain

$$|m| \leqslant |n|^{(\log m)/(\log n)} \quad \text{or} \quad |m|^{1/(\log m)} \leqslant |n|^{1/(\log n)}.$$

By interchanging the roles of m and n we see that for all integers m and n which are greater than 1,

$$|m|^{1/(\log m)} = |n|^{1/(\log n)} = e^c$$

for some positive real number c. Thus for all $n \in J$

$$|n| = \exp\left(c \log \text{(absolute value of } n)\right)$$

$$= \text{(absolute value of } n)^c.$$

Therefore, $|\;|$ is equivalent to ordinary absolute value. ‖

Let k be an arbitrary field and consider the field $k(x)$ of rational functions over k. Choose once and for all a real number d with $0 < d < 1$. If $f(x)/g(x)$ is an arbitrary nonzero element of $k(x)$ we set

$$|f(x)/g(x)|_\infty = d^{\deg g(x) - \deg f(x)}.$$

Further we set $|0|_\infty = 0$. This is a well-defined real-valued function on $k(x)$; we shall show that it is a valuation on $k(x)$. Clearly (i) and (ii) in the definition of valuation are satisfied, as is (iii) when at least one of the elements of $k(x)$ involved is zero. Let $f(x)/g(x)$ and $r(x)/q(x)$ be nonzero elements of $k(x)$ with $\deg g(x) - \deg f(x) \leqslant \deg q(x) - \deg r(x)$. Then

$$\left|\frac{f(x)}{g(x)} + \frac{r(x)}{q(x)}\right|_\infty = \left|\frac{f(x)q(x) + r(x)g(x)}{g(x)q(x)}\right|_\infty$$

$$= d^{\deg g(x)q(x) - \deg (f(x)q(x) + r(x)g(x))}.$$

Now $\deg (f(x)q(x) + r(x)g(x)) \leqslant \max (\deg f(x)q(x), \deg r(x)g(x))$, and so the exponent on d is either

$$\geqslant \deg g(x)q(x) - \deg f(x)q(x) = \deg g(x) - \deg f(x)$$

or

$$\geqslant \deg g(x)q(x) - \deg r(x)g(x) = \deg q(x) - \deg r(x)$$

$$\geqslant \deg g(x) - \deg f(x).$$

Thus in either case

$$\left| \frac{f(x)}{g(x)} + \frac{r(x)}{q(x)} \right|_\infty \leq d^{\deg g(x) - \deg f(x)}$$

$$= \max \left(\left| \frac{f(x)}{g(x)} \right|_\infty, \left| \frac{r(x)}{q(x)} \right|_\infty \right).$$

Therefore, $| \ |_\infty$ is a non-Archimedean valuation on $k(x)$ and is clearly trivial on k.

Let $p(x)$ be a nonconstant irreducible polynomial in $k[x]$. If $f(x)/g(x)$ is an arbitrary nonzero element of $k(x)$ we can write.

$$\frac{f(x)}{g(x)} = p(x)^n \frac{u(x)}{v(x)},$$

where $u(x)$ and $v(x)$ are relatively prime elements of $k[x]$, neither of which is divisible by $p(x)$. We set

$$\left| \frac{f(x)}{g(x)} \right|_{p(x)} = d^n, \qquad |0|_{p(x)} = 0.$$

We can show that $| \ |_{p(x)}$ is a non-Archimedean valuation on $k(x)$, which is done in the same way as it was shown that $| \ |_p$ is a non-Archimedean valuation on Q. Clearly $| \ |_{p(x)}$ is trivial on k.

Note that if we replace d by d', where $0 < d' < 1$, in the definitions of $| \ |_\infty$ and $| \ |_{p(x)}$ we obtain valuations on $k(x)$ which are equivalent to $| \ |_\infty$ and $| \ |_{p(x)}$, respectively.

THEOREM 3. *Let $k(x)$ be the field of rational functions over a field k. Let $| \ |$ be a nontrivial valuation on $k(x)$ which is trivial on k. Then $| \ |$ is equivalent to either $| \ |_\infty$ or $| \ |_{p(x)}$ for some irreducible polynomial $p(x) \in k[x]$.*

Proof. Since $k(x)$ is the field of quotients of $k[x]$ it is sufficient to show that $| \ |$ is equivalent to $| \ |_\infty$ or to some $| \ |_{p(x)}$ on $k[x]$. Since $| \ |$ is trivial on k it is non-Archimedean on $k[x]$ by the lemma proved above. We consider two cases.

Case 1. $|x| > 1$. Let $f(x) \in k[x]$,

$$f(x) = a_0 + a_1 x + \cdots + a_n x^n, \qquad a_n \neq 0.$$

Then $|a_i x^i| < |a_j x^j|$ for $i < j$ and so

$$|f(x)| = |x^n| = |x|^n = d^{-\deg f(x)}, \qquad d = |x|^{-1}.$$

Hence $| \ |$ is equivalent to $| \ |_\infty$.

Case 2. $|x| \leq 1$. Then $|f(x)| \leq 1$ for all $f(x) \in k[x]$. Let

$$I = \{ f(x) \in k[x] \, | \, |f(x)| < 1 \}.$$

Then I is a prime ideal of $k[x]$ so that $I = (p(x))$ where $p(x)$ is an irreducible polynomial in $k[x]$. If $f(x) \in k[x]$ and if $p(x)$ does not divide $f(x)$ then $|f(x)| = 1$. Now let $|p(x)| = d$ (then $0 < d < 1$) and let $f(x) = p(x)^n g(x)$ where $g(x)$ is not divisible by $p(x)$. Then $|f(x)| = |p(x)|^n = d^n$ and therefore $|\ |$ is equivalent to $|\ |_{p(x)}$. ‖

3. Complete fields and completions

Let $|\ |$ be an arbitrary valuation on a field k. A sequence $\{a_n\}$ of elements of k (n will always run from 0 to ∞) is said to be *convergent* if there is an $a \in k$ such that given $\varepsilon > 0$ there is an integer $n_0 = n_0(\varepsilon)$ such that $|a_n - a| < \varepsilon$ whenever $n > n_0$. In this case we say that the sequence *converges* to a, and we write

$$\lim_{n \to \infty} a_n = a.$$

A sequence $\{a_n\}$ of elements of k is called a *null sequence* if $\lim_{n \to \infty} a_n = 0$.

A sequence $\{a_n\}$ of elements of k is called a *Cauchy sequence* if for any $\varepsilon > 0$ there is an integer $n_0 = n_0(\varepsilon)$ such that $|a_m - a_n| < \varepsilon$ whenever m, $n > n_0$. Every convergent sequence of elements of k is a Cauchy sequence. This statement, like many others in this section, will not be proved. Its proof is exactly the same as the proof of the corresponding statement concerning sequences of real numbers.

Let $\mathfrak{S}$ be the set consisting of all Cauchy sequences of elements of k. With each pair of sequences $\{a_n\}$ and $\{b_n\}$ we can associate another pair of sequences defined by

$$\{a_n\} + \{b_n\} = \{a_n + b_n\}$$

and

$$\{a_n\}\{b_n\} = \{a_n b_n\}.$$

These new sequences, called the *sum* and *product* of $\{a_n\}$ and $\{b_n\}$, respectively, are Cauchy sequences if $\{a_n\}$ and $\{b_n\}$ are Cauchy sequences. Let $\mathfrak{N}$ be the set of all null sequences of elements of k. Sums and products of null sequences are null sequences.

PROPOSITION. *With respect to the operations of addition and multiplication defined above, $\mathfrak{S}$ is a commutative ring and $\mathfrak{N}$ is a maximal ideal of $\mathfrak{S}$.*

A *valuated field* is a field k together with a valuation $|\ |$ on k. The composite object $(k, |\ |)$ will generally be denoted simply by k. When we wish to emphasize that $|\ |$ is the valuation of the valuated field k, we shall write $|\ |_k$. If $|\ |_k$ is non-Archimedean then k is called a *non-Archimedean valuated field*. A valuated field K is called an *extension* of the valuated field k if the field K is an extension of the field k and if $|a|_K = |a|_k$ for all $a \in k$. In this case we refer to $|\ |_k$ as the *restriction* of $|\ |_K$ to k and to $|\ |_K$ as a *prolongation* of $|\ |_k$ to K. If K and L are valuated fields then an isomorphism σ from K into L is called *analytic* if $|a|_K = |\sigma(a)|_L$ for all $a \in K$.

DEFINITION. A valuated field k is said to be *complete* if every Cauchy sequence of elements of k is a convergent sequence.

There are valuated fields which are not complete. For example, the rational field Q with ordinary absolute value is not complete, as is well-known. Moreover, there are non-Archimedean valuated fields which are not complete, such as Q with any p-adic valuation. We wish to investigate the possibility of embedding a valuated field in a complete valuated field which is the smallest in some sense. As we shall see, this can be done by mimicking the process by which the field Q with ordinary absolute value is embedded in the field of real numbers with ordinary absolute value.

DEFINITION. Let k be a valuated field. A valuated field K is called a *completion* of k if it satisfies the following conditions:
  (i) K is an extension of k,
  (ii) K is complete,
  (iii) every element of K is the limit of some Cauchy sequence of elements of k.

THEOREM 4. *Let k be a valuated field. Then*
  (a) *there exists a completion of k,*
  (b) *if K is a completion of k and if L is a complete valuated field which is an extension of k then there is an analytic k-isomorphism from K onto some subfield of L, and in particular,*
  (c) *any two completions of k are analytically k-isomorphic.*

We shall do little more than outline the proof of this theorem. Consider the ring $\mathfrak{S}$ of Cauchy sequences of elements of k and its maximal ideal of null sequences $\mathfrak{N}$. The residue class ring $\mathfrak{S}/\mathfrak{N}$ is a field K. We define a mapping ϕ from k into K by $\phi(a) = \{a\} + \mathfrak{N}$ where $\{a\}$ is the sequence having each of its terms equal to a. The mapping ϕ is an isomorphism. We identify each element of k with its image under ϕ so that the field K is an extension of the field k.

If $\{a_n\}$ is a Cauchy sequence of elements of k then $\{|a_n|_k\}$ is a Cauchy sequence of real numbers. Since the real numbers are complete with respect to ordinary absolute value, this sequence converges. We set

$$|\{a_n\} + \mathfrak{N}|_K = \lim_{n \to \infty} |a_n|_k.$$

This is a well-defined real-valued function on K which is actually a valuation on K. Since

$$|\{a\} + \mathfrak{N}|_K = \lim_{n \to \infty} |a|_k = |a|_k,$$

the resulting valuated field K is an extension of the valuated field k. Furthermore, the valuated field K is complete and every element of K is the limit of a Cauchy sequence of elements of k. In fact,

$$\{a_n\} + \mathfrak{N} = \lim_{n \to \infty} a_n.$$

Thus K is a completion of k.

The analytic k-isomorphism of (b) is obtained in the following manner. If $\{a_n\}$ is a Cauchy sequence of elements of k we simply map $\lim_{n \to \infty} a_n$ in K onto $\lim_{n \to \infty} a_n$ in L. If K and L are both completions of k, the k-isomorphism so defined will map K onto L.

4. Value groups and residue class fields

Let k be a valuated field. For all a, $b \in k$ we have $|ab| = |a| \, |b|$ and therefore the mapping $a \to |a|$ is a homomorphism from k^* into the multiplicative group of positive real numbers. Therefore,

$$V_k = \{|a| \mid a \in k^*\}$$

is a group which is called the *value group* of the valuated field k. If K is a valuated field which is an extension of k, then $V_k \subseteq V_K$.

Throughout this section we shall assume that the valuated field k is non-Archimedean.

THEOREM 5. *If K is a completion of the valuated field k then $V_k = V_K$.*

Proof. Let $a \in K^*$: we must show that $|a| \in V_k$. Since K is a completion of k we have $a = \lim_{n \to \infty} a_n$ where $\{a_n\}$ is a Cauchy sequence of elements of k. Given $\varepsilon > 0$, chosen so that $\varepsilon < |a|$, there is an integer $n_0 = n_0(\varepsilon)$ such that $|a - a_n| < \varepsilon$ whenever $n > n_0$. Then, for all $n > n_0$, $|a_n| = |(a_n - a) + a| = \max(|a_n - a|, |a|) = |a|$ and so $|a| \in V_k$. ‖

Let

$$\mathfrak{o} = \{a \mid a \in k \text{ and } |a| \leqslant 1\}$$

and

$$\mathfrak{p} = \{a \mid a \in k \text{ and } |a| < 1\}.$$

Then $\mathfrak{o}$ is a subring of k, for if a, $b \in \mathfrak{o}$ then $|a - b| \leqslant \max(|a|, |b|) \leqslant 1$ and $|ab| = |a| \, |b| \leqslant 1$. Since $\mathfrak{o}$ is a subring of a field it is an integral domain and k is a field of quotients of $\mathfrak{o}$. To see this let $a \in k$: if $|a| \leqslant 1$ then $a \in \mathfrak{o}$; if $|a| > 1$ then $a^{-1} \in \mathfrak{o}$ and we have $a = 1/a^{-1}$. Furthermore, $\mathfrak{p}$ is an ideal of $\mathfrak{o}$, for if a, $b \in \mathfrak{p}$ and $c \in \mathfrak{o}$ then $|a - b| \leqslant \max(|a|, |b|) < 1$ and $|ac| = |a| \, |c| < 1$. It is a maximal ideal (and therefore a prime ideal) of $\mathfrak{o}$, for if $\mathfrak{a}$ is an ideal of $\mathfrak{o}$ such that $\mathfrak{a} \nsubseteq \mathfrak{p}$ then there is an element $u \in \mathfrak{a}$ such that $|u| = 1$; then $u^{-1} \in \mathfrak{o}$ and so $uu^{-1} = 1 \in \mathfrak{a}$ which means that $\mathfrak{a} = \mathfrak{o}$. The ring $\mathfrak{o}$ is called the *ring of integers* of the valuated field k. If U is the set of all units of $\mathfrak{o}$, that is, if $U = \{u \in k| \ |u| = 1\}$, then U is a group with respect to multiplication which is called the *group of units* of k. The ideal $\mathfrak{p}$ of $\mathfrak{o}$ is precisely the set of nonunits of $\mathfrak{o}$. We note that every ideal of $\mathfrak{o}$ other than $\mathfrak{o}$ itself is contained in $\mathfrak{p}$.

Since $\mathfrak{p}$ is a maximal ideal of $\mathfrak{o}$ the residue class ring $\mathfrak{o}/\mathfrak{p}$ is a field which we denote by $\bar{k}$ and call the *residue class field* of the valuated field k.

THEOREM 6. *Let K be an extension of the valuated field k. Then there is an isomorphism from $\bar{k}$ onto a subfield of $\bar{K}$.*

Proof. Let $\mathfrak{o}$ be the ring of integers of k and $\mathfrak{p}$ its ideal of nonunits and let $\mathfrak{O}$ be the ring of integers of K and $\mathfrak{P}$ its ideal of nonunits. Then it is clear that $\mathfrak{o} \subseteq \mathfrak{O}$ and $\mathfrak{p} \subseteq \mathfrak{P}$. Hence if we define ϕ from $\bar{k}$ into $\bar{K}$ by $\phi(a + \mathfrak{p}) = a + \mathfrak{P}$, ϕ is well-defined. Furthermore, ϕ is a homomorphism and since the elements in $\mathfrak{o}$ but not in $\mathfrak{p}$ are also not in $\mathfrak{P}$, ϕ is not the trivial homomorphism. Hence ϕ is an isomorphism. $\|$

From now on, when we deal with the situation described in Theorem 6, we shall identify $a + \mathfrak{p}$ and $a + \mathfrak{P}$, where $a \in \mathfrak{o}$, and so consider $\bar{K}$ as an extension of $\bar{k}$.

THEOREM 7. *Let K be a completion of a valuated field k. Then $\bar{K} = \bar{k}$.*

Proof. In the light of the proof of Theorem 6, we must show that if $a \in \mathfrak{O}$ then there is an element $b \in \mathfrak{o}$ such that $b + \mathfrak{P} = a + \mathfrak{P}$, that is, $|b - a| < 1$. Let $a = \lim_{n \to \infty} a_n$ where $\{a_n\}$ is a Cauchy sequence of elements of k. Then there is an integer n_0 such that $|a - a_n| < 1$ whenever $n > n_0$. Furthermore, for such n we have $|a_n| = |(a_n - a) + a| \leqslant \max(|a_n - a|, |a|) \leqslant 1$ and so $a_n \in \mathfrak{o}$. Thus we may take for b any a_n with $n > n_0$. $\|$

In the remainder of this section we shall determine the residue class fields of certain valuated fields. The completion of Q with respect to the p-adic valuation $|\ |_p$ is called the *field of p-adic numbers* and we shall denote this field by Q_p. Elements of the ring of integers of Q_p are called *p-adic integers*. In order to determine $\bar{Q}_p$ we note that by Theorem 7 this is the same as the residue class field of the valuated field $(Q, |\ |_p)$. Let $\mathfrak{o}$ be the ring of integers of $(Q, |\ |_p)$ and $\mathfrak{p}$ its ideal of nonunits. Then

$$\mathfrak{o} = \{a/b \mid a \text{ and } b \text{ are integers and } p \nmid b\}$$

and

$$\mathfrak{p} = \{a/b \mid a \text{ and } b \text{ are integers, } p \mid a, \text{ and } p \nmid b\}.$$

Define a mapping ϕ from $\mathfrak{o}$ into $GF(p)$ by

$$\phi(a/b) = [a][b]^{-1}.$$

This is obviously a well-defined mapping and is a homomorphism. Since

$\phi(a/1) = [a]$, ϕ is *onto*. We have a/b in the kernel of ϕ if and only if $[a] = [0]$, that is, if and only if $p \mid a$, that is, if and only if $a/b \in \mathfrak{p}$. Therefore,

$$\bar{Q}_{\mathfrak{p}} = \mathfrak{o}/\mathfrak{p} \cong GF(p).$$

Now consider the field $F(x)$ of rational functions over an arbitrary field F. Let $p(x)$ be an irreducible polynomial in $F[x]$. We shall show that the residue class field of the valuated field $(F(x), \mid \mid_{p(x)})$ is isomorphic to $F(a)$ where a is a root of $p(x)$.

Let $\mathfrak{o}$ be the ring of integers of this valuated field and let $\mathfrak{p}$ be its ideal of nonunits. Then

$$\mathfrak{o} = \left\{ \frac{f(x)}{g(x)} \middle| f(x), g(x) \in F[x] \quad \text{and} \quad p(x) \nmid g(x) \right\}$$

and

$$\mathfrak{p} = \left\{ \frac{f(x)}{g(x)} \middle| f(x), g(x) \in F[x], p(x) \nmid g(x), \quad \text{and} \quad p(x) \mid f(x) \right\}.$$

Let $f(x)/g(x) \in \mathfrak{o}$. Then $p(x) \nmid g(x)$ and so there is a polynomial $h(x) \in F[x]$ such that $g(x)h(x) \equiv 1 \pmod{p(x)}$. Define ϕ from $\mathfrak{o}$ into $F[x]/(p(x))$ by $\phi(f(x)/g(x)) = f(x)h(x) + (p(x))$. We show first that ϕ is well-defined. Suppose that $f(x)/g(x) = f_1(x)/g_1(x)$ and let $g_1(x)h_1(x) \equiv 1 \pmod{p(x)}$. Then $g(x)h(x) \equiv g_1(x)h_1(x) \pmod{p(x)}$ and so $g(x)f(x)h(x) \equiv g_1(x)f(x)h_1(x) \pmod{p(x)}$. But $f(x) = g(x)f_1(x)/g_1(x)$ and so $g(x)f(x)h(x) \equiv g(x)f_1(x)h_1(x) \pmod{p(x)}$. Since $p(x) \nmid g(x)$ we conclude that $f(x)h(x) \equiv f_1(x)h_1(x) \pmod{p(x)}$, or $f(x)h(x) + (p(x)) = f_1(x)h_1(x) + (p(x))$. Thus ϕ is well-defined and it is certainly a homomorphism. It is *onto* since $\phi(f(x)) = f(x) + (p(x))$. Furthermore, the kernel of ϕ is precisely $\mathfrak{p}$ and so $\overline{F(x)} \cong F[x]/(p(x))$. However, it was shown in Chapter 1 that this field is $F(a)$ where a is a root of $p(x)$.

5. Prolongations of valuations

Let k be a valuated field and let K be an extension of the field k. In this section, and in the two that follow, we shall be concerned with the following problem: is there a valuation on K which is a prolongation of the valuation on k, and if there is at least one such prolongation, how many prolongations are there. We shall be concerned only with the case in which K is an algebraic extension of k. In this section we shall show that there is always at least one prolongation and we shall give a preliminary discussion of the possible uniqueness of the prolongation. In the next section we shall thoroughly discuss the case in which there is a unique prolongation, and in the final section of this chapter we shall consider the general question of the number of prolongations.

Let k be a non-Archimedean valuated field with valuation $|\ |$. If $|\ |$ is the trivial valuation on k then it certainly has a prolongation to any extension of k, namely, the trivial valuation on that extension. Therefore, we shall always assume that $|\ |$ is nontrivial. The valuation $|\ |$, restricted to k^*, is a homomorphism from k^* onto the value group V_k. This is a subgroup of the multiplicative group of positive real numbers. Hence it is an example of what we shall call an *ordered Abelian group*, that is, it has a total ordering (in this case, the natural ordering $\leqslant$ of the real numbers) such that if a, b, and c are elements of the group and if $a \leqslant b$ then $ac \leqslant bc$. The kernel of the homomorphism from k^* onto V_k is the group U of units of k and so there is an isomorphism ϕ from k^*/U onto V_k which is given by $\phi(aU) = |a|$. We can make k^*/U into an ordered Abelian group by requiring that ϕ be an order-preserving isomorphism, that is, we write $aU \leqslant bU$ if and only if $\phi(aU) \leqslant \phi(bU)$. This is equivalent to $|a| \leqslant |b|$, or $|ab^{-1}| \leqslant 1$, or $ab^{-1} \in \mathfrak{o}$, where $\mathfrak{o}$ is the ring of integers of k.

THEOREM 8. *If K is an extension of the field k then there is a subring $\mathfrak{O}$ of K with the following properties:*

(i) $\mathfrak{o} \subseteq \mathfrak{O}$ *but* $\mathfrak{O} \neq K$,

(ii) *if* $a \in K$ *and* $a \notin \mathfrak{O}$ *then* $a^{-1} \in \mathfrak{O}$.

Proof. Let $\mathfrak{p}$ be the ideal of nonunits of $\mathfrak{o}$, and let $\mathfrak{S}$ be the set of all subrings $\mathfrak{a}$ of K such that $\mathfrak{o} \subseteq \mathfrak{a}$ and $\mathfrak{pa} \neq \mathfrak{a}$. Here $\mathfrak{pa}$ is the ideal of $\mathfrak{a}$ generated by $\mathfrak{p}$: we have

$$\mathfrak{pa} = \{\text{finite sums } \textstyle\sum c_i a_i \mid c_i \in \mathfrak{p} \text{ and } a_i \in \mathfrak{a}\}.$$

The set $\mathfrak{S}$ is not empty since $\mathfrak{o} \in \mathfrak{S}$, and we note that $K \notin \mathfrak{S}$. We make $\mathfrak{S}$ into a partially ordered set by set inclusion and we consider the totally ordered subset $\{\mathfrak{a}_\nu \mid \nu \in N\}$ of $\mathfrak{S}$. We shall show that this totally ordered subset of $\mathfrak{S}$ has an upper bound in $\mathfrak{S}$. Let $\mathfrak{a}$ be the union of the $\mathfrak{a}_\nu$. Then $\mathfrak{a}$ is a subring of K which contains each $\mathfrak{a}_\nu$ and so contains $\mathfrak{o}$. If we show that $\mathfrak{pa} \neq \mathfrak{a}$ then $\mathfrak{a} \in \mathfrak{S}$ and it is clearly an upper bound of the totally ordered set. If $\mathfrak{pa} = \mathfrak{a}$ then $1 \in \mathfrak{pa}$ so that $1 = c_1 a_1 + \cdots + c_n a_n$ where $c_1, \ldots, c_n \in \mathfrak{p}$ and $a_1, \ldots, a_n \in \mathfrak{a}$. Each a_i is in some $\mathfrak{a}_\nu$ and so there is an element $\nu \in N$ such that $a_i \in \mathfrak{a}_\nu$ for $i = 1, \ldots, n$. Then $1 \in \mathfrak{pa}_\nu$ and so $\mathfrak{pa}_\nu = \mathfrak{a}_\nu$, which is not true. This contradiction leads us to the conclusion that $\mathfrak{pa} \neq \mathfrak{a}$. Therefore, by Zorn's lemma, the set $\mathfrak{S}$ has a maximal element $\mathfrak{O}$. Since $\mathfrak{O} \in \mathfrak{S}$ we have $\mathfrak{pO} \neq \mathfrak{O}$ and so $\mathfrak{O} \neq K$. Also $\mathfrak{o} \subseteq \mathfrak{O}$.

Consider the set $A = \{a \in \mathfrak{O} \mid a - 1 \in \mathfrak{pO}\}$. If a, $b \in A$ then $ab - 1 = a(b-1) + (a-1) \in \mathfrak{pO}$ and therefore $ab \in A$. Clearly $0 \notin A$. If we set $\mathfrak{O}' = \{ab^{-1} \mid a \in \mathfrak{O} \text{ and } b \in A\}$ then it is easy to verify that $\mathfrak{O}'$ is a subring of K and $\mathfrak{O} \subseteq \mathfrak{O}'$. We shall show that $\mathfrak{O} = \mathfrak{O}'$. If $\mathfrak{O} \subset \mathfrak{O}'$ then by the maximality of $\mathfrak{O}$ in $\mathfrak{S}$ we have $\mathfrak{O}' \notin \mathfrak{S}$ and so $\mathfrak{pO}' = \mathfrak{O}'$. This means that there are elements $c_1, \ldots, c_n \in \mathfrak{p}$, $a_1, \ldots, a_n \in \mathfrak{D}$, and $b_1, \ldots, b_n \in A$ such that $1 = c_1 a_1 b_1^{-1} + \cdots + c_n a_n b_n^{-1}$. If we let $b = b_1 \cdots b_n$ then $b \in A$ and $b = c_1 a_1' + \cdots + c_n a_n'$ where for each i, $a_i' = a_i b b_i^{-1} \in \mathfrak{D}$. Thus $b \in \mathfrak{p}\mathfrak{D}$. However, $b - 1 \in \mathfrak{p}\mathfrak{D}$ since $b \in A$, and therefore $1 \in \mathfrak{p}\mathfrak{D}$ which implies that $\mathfrak{p}\mathfrak{D} = \mathfrak{D}$, a contradiction. Thus we must have $\mathfrak{D} = \mathfrak{D}'$ and, therefore, each element of A has its inverse in $\mathfrak{D}$.

Now let $a \in K$ and assume that $a \notin \mathfrak{D}$. Then $\mathfrak{D} \subset \mathfrak{D}[a]$ and $\mathfrak{D}[a] \notin \mathfrak{S}$ which means that $\mathfrak{p}\mathfrak{D}[a] = \mathfrak{D}[a]$. Hence there are elements $c_1, \ldots, c_n \in \mathfrak{p}$ and $b_1, \ldots, b_n \in \mathfrak{D}[a]$ such that $1 = c_1 b_1 + \cdots + c_n b_n$. Each b_i has the form $r(a)$ where $r(x) \in \mathfrak{D}[x]$ and as a result we have $1 = d_0 + d_1 a + \cdots + d_m a^m$ for some m with $d_0, d_1, \ldots, d_m \in \mathfrak{p}\mathfrak{D}$. Since $(1 - d_0) - 1 = -d_0 \in \mathfrak{p}\mathfrak{D}$ we have $1 - d_0 \in A$ and therefore $1 = d_1' a + \cdots + d_m' a^m$ where for each i, $d_i' = d_i(1 - d_0)^{-1} \in \mathfrak{p}\mathfrak{D}$. Now assume that m is the smallest possible integer such that 1 can be written in this way. We would like to show that $a^{-1} \in \mathfrak{D}$. Suppose that $a^{-1} \notin \mathfrak{D}$. Then we can repeat the discussion above with a replaced by a^{-1}, and we let t be the smallest possible integer such that $1 = e_1 a^{-1} + \cdots + e_t (a^{-1})^t$ with $e_1, \ldots, e_t \in \mathfrak{p}\mathfrak{D}$. Then $a^t = e_1 a^{t-1} + \cdots + e_t$. If $t \leqslant m$ then $1 = d_1' a + \cdots + d'_{m-1} a^{m-1} + d_m' a^{m-t}(e_1 a^{t-1} + \cdots + e_t)$ and the largest power of a that appears in this expression is the $(m-1)$st. This contradicts our choice of m. If $m \leqslant t$ we are led in the same manner to a contradiction of our choice of t. Thus we must conclude that $a^{-1} \in \mathfrak{D}$. Therefore, $\mathfrak{D}$ meets the requirements of the theorem. ‖

Let k be a non-Archimedean valued field and let K be an extension of the field k. Let $\mathfrak{o}$ be the ring of integers of k, $\mathfrak{p}$ the ideal of nonunits of $\mathfrak{o}$, and let U_k be the group of units of k. Let $\mathfrak{D}$ be a subring of K which satisfies (i) and (ii) of Theorem 8. Let $\mathfrak{P}$ be the set of nonunits of $\mathfrak{D}$ and U_K the group of units of the ring $\mathfrak{D}$. We shall show that $\mathfrak{P}$ is an ideal of $\mathfrak{D}$. It is clear that the product of a nonunit of $\mathfrak{D}$ and an arbitrary element of $\mathfrak{D}$ is a nonunit of $\mathfrak{D}$. We must also show that if $a, b \in \mathfrak{P}$ then $a - b \in \mathfrak{P}$. If either $a = 0$ or $b = 0$ then $a - b$ is certainly in $\mathfrak{P}$. Hence we assume that $a \neq 0$ and $b \neq 0$. Then either $ab^{-1} \in \mathfrak{D}$ or $ba^{-1} \in \mathfrak{D}$. If $ab^{-1} \in \mathfrak{D}$ then $ab^{-1} - 1 \in \mathfrak{D}$ and so $a - b = (ab^{-1} - 1)b \in \mathfrak{P}$. If $ba^{-1} \in \mathfrak{D}$ then $ba^{-1} - 1 \in \mathfrak{D}$ and $a - b = -(ba^{-1} - 1)a \in \mathfrak{P}$.

Since U_K is a subgroup of K^* we may consider the group K^*/U_K. We define a mapping from k^*/U_k into K^*/U_K as follows: if $a \in k^*$ we map aU_k onto aU_K. This mapping is certainly a homomorphism and since $U_k = \mathfrak{o} \cap U_K = k^* \cap U_k$, it is an isomorphism. We shall identify each element of k^*/U_k with its image in K^*/U_K under this mapping and so regard k^*/U_k as a subgroup of K^*/U_K. We now introduce an ordering into K^*/U_K by defining $aU_K \leqslant bU_K$ to mean that $ab^{-1} \in \mathfrak{D}$. This is a total ordering because, given $a, b \in K^*$, either $ab^{-1} \in \mathfrak{D}$ or $ba^{-1} \in \mathfrak{D}$, so that either $aU_K \leqslant bU_K$ or $bU_K \leqslant aU_K$. With this ordering K^*/U_K becomes an ordered Abelian group. Furthermore, the ordering on K^*/U_K induces on the subgroup k^*/U_k exactly the same ordering that we discussed earlier in this section.

There is an important property of the ordering on K^*/U_K which we will require below. Let $a, b \in K^*$ and suppose that $a + b \neq 0$. Further suppose that $aU_K \leqslant bU_K$. Then $ab^{-1} \in \mathfrak{D}$ and therefore $ab^{-1} + 1 \in \mathfrak{D}$: hence $(a + b)U_K = (ab^{-1} + 1)bU_K \leqslant bU_K = \max (aU_K, bU_K)$. In fact, if $aU_K < bU_K$ then $(a + b)U_K = bU_K$. For assume that $(a + b)U_K < bU_K$. Then $bU_K = ((a + b) - a)U_K \leqslant \max ((a + b)U_K, aU_K) < bU_K$, a contradiction. It follows in like manner that if $a_1, \ldots, a_n \in K^*$, if $a_1 + \cdots + a_n \neq 0$ and $a_2 + \cdots + a_n \neq 0$, and if $a_i U_K < a_1 U_K$ for $i = 2, \ldots, n$, then $(a_1 + \cdots + a_n)U_K = a_1 U_K$.

LEMMA.   *Let G be an ordered Abelian group and assume that G has a subgroup H with the following properties:*
 (a) *there is an order-preserving isomorphism ϕ from H into the multiplicative group of positive real numbers,*
 (b) *if $a \in G$, there is a positive integer n such that $a^n \in H$.*
Then there is an order-preserving isomorphism ψ from G into the multiplicative group of positive real numbers such that $\psi(a) = \phi(a)$ for all $a \in H$.

Proof.   Let $a \in G$ and let n be a positive integer such that $a^n \in H$. Then $\phi(a^n)$ has a unique positive real nth root $\phi(a^n)^{1/n}$. We set $\psi(a) = \phi(a^n)^{1/n}$. If we also have $a^m \in H$ for the positive integer m, then $a^{mn} \in H$ and $\phi(a^{mn}) = \phi(a^m)^n = \phi(a^n)^m$. Hence $\phi(a^m)^{1/m} = \phi(a^n)^{1/n}$ and consequently ψ is a well-defined mapping from G into the positive real numbers. If $a \in H$ then $\psi(a) = \phi(a)$. Now let $a, b \in G$ and let n be a positive integer such that $a^n \in H$ and $b^n \in H$. Since G is Abelian, $(ab)^n = a^n b^n \in H$, and we have $\psi(ab) = \phi(a^n b^n)^{1/n} = \phi(a^n)^{1/n} \phi(b^n)^{1/n} = \psi(a)\psi(b)$. Thus ψ is a homomorphism from G into the multiplicative group of positive real numbers. Suppose that $\psi(a) = 1$ for some $a \in G$. Then for some positive integer n we have $a^n \in H$ and $\phi(a^n) = 1$. Since ϕ is an isomorphism this means that $a^n = e$, the identity element of G. If $a \leqslant e$ then $e \leqslant a^{-1}$ and $(a^{-1})^n = e$, so that we may assume in the first place that $e \leqslant a$. If $e < a$ then $a < a^2$, $a^2 < a^3$, and so on, until we finally have $e < a^n$, so that we must have $a = e$. Therefore ψ is an isomorphism. Finally, assume that $a, b \in G$ and that $a \leqslant b$. Let n be a positive integer such that $a^n \in H$ and $b^n \in H$. Then $a^n \leqslant b^n$ and $\phi(a^n) \leqslant \phi(b^n)$. Therefore, $\psi(a) \leqslant \psi(b)$ and ψ is order-preserving. ‖

We can now prove the principal result of this section, which completely settles the question of the existence of prolongations of non-Archimedean valuations on a field k to algebraic extensions of k.

THEOREM 9.   *Let k be a non-Archimedean valued field with valuation $| \ |$. If K is an algebraic extension of k then there is a prolongation of $| \ |$ to a valuation on K.*

Proof. We continue to use the notation we have used above. Let $a \in K^*$ and let Irr $(k,a) = c_0 + c_1 x + \cdots + c_n x^n$ $(c_n = 1)$. Then $c_0 + c_1 a + \cdots + c_n a^n = 0$ and $c_0 \neq 0$. We now have to consider two possibilities. First, it may happen that there are two integers i and j with $0 \leqslant i < j \leqslant n$ such that $c_i \neq 0$, $c_j \neq 0$, and $c_i a^i U_K = c_j a^j U_K$. Then $(aU_K)^{j-i} = a^{j-i} U_K = c_i c_j^{-1} U_K \in k^*/U_k$. On the other hand, it may happen that for all such i and j, $c_i a^i U_K \neq c_j a^j U_K$. Then, since $c_1 a + \cdots + c_n a^n = -c_0 \neq 0$, and since no subset of the set $\{c_1 a, \ldots, c_n a^n\}$ containing at least one nonzero element has the sum of its elements equal to zero, there is an integer i with $1 \leqslant i \leqslant n$ such that $c_i \neq 0$ and $(-c_0)U_K = c_i a^i U_K$. Then $(aU_K)^i = a^i U_K = (-c_0)c_i^{-1} U_K \in k^*/U_k$. Thus, for each $aU_K \in K^*/U_K$ there is a positive integer m such that $(aU_K)^m \in k^*/U_k$.

Now, by the lemma proved above, there is an order-preserving isomorphism ψ from K^*/U_K into the multiplicative group of positive real numbers such that $\psi(aU_K) = |a|$ for all $a \in k^*$. We define the real-valued function $|\ |_K$ on K by

$$|a|_K = \psi(aU_K) \quad \text{if } a \in K^* \quad \text{and} \quad |0|_K = 0.$$

The mapping $a \to |a|_K$, for $a \in K^*$, is the composite of the natural homomorphism from K^* onto K^*/U_K and ψ. Hence $|ab|_K = |a|_K |b|_K$, and this is certainly true when $a = 0$ or $b = 0$. Now let a, $b \in K$. If $a = 0$ then $|a + b|_K = |b|_K = \max (|a|_K, |b|_K)$, and likewise when $b = 0$. Assume that $a \neq 0$ and $b \neq 0$. If $a + b = 0$ then $|a + b|_K = 0 < \max (|a|_K, |b|_K)$. Suppose then that $a + b \neq 0$ and that $|a|_K \leqslant |b|_K$. Thus $\psi(aU_K) \leqslant \psi(bU_K)$, $aU_K \leqslant bU_K$. But then $(a + b)U_K \leqslant bU_K$ so that $\psi((a + b)U_K) \leqslant \psi(bU_K)$ and we have $|a + b|_K \leqslant |b|_K = \max (|a|_K, |b|_K)$. Therefore $|\ |_K$ is a valuation on K and if $a \in k$ we have $|a|_K = |a|$ so that $|\ |_K$ is a prolongation of $|\ |$. ∥

Note that $a \in \mathfrak{O}$ if and only if $aU_K \leqslant U_K$, that is, if and only if $|a|_K \leqslant 1$. Thus $\mathfrak{O}$ is the ring of integers of the valuated field $(K, |\ |_K)$.

As we shall see later, there may be several prolongations of a valuation on a field k to valuations on an extension of k. We shall now consider the question of when there is only one prolongation to a valuation on an algebraic extension of k.

THEOREM 10. *Let k be a complete valuated field with valuation $|\ |$. Let K be a finite extension of k and let $|\ |_K$ be a prolongation of $|\ |$ to a valuation on K. Then K is complete with respect to $|\ |_K$.*

Proof. Let $[K: k] = n$ and let $a_1, \ldots, a_n$ be a basis of K over k. Let $\{b_r\}$ be a Cauchy sequence of elements of K. Then for each r,

$$b_r = c_{r1} a_1 + \cdots + c_{rn} a_n, \tag{1}$$

where $c_{ri} \in k$ for all r and $i = 1, \ldots, n$. Assume that for each i, $\{c_{ri}\}$ is a Cauchy sequence of elements of k and let $c_i = \lim_{r \to \infty} c_{ri}$. For a given $\varepsilon > 0$

let $r_0 = r_0(\varepsilon)$ be such that $|c_i - c_{ri}| < \varepsilon/nM$ for each i whenever $r > r_0$, where $M = \max_{1 \leqslant i \leqslant n} |a_i|_K$. Then, for $r > r_0$, we have

$$\left| \sum_{i=1}^{n} c_i a_i - b_r \right|_K = \left| \sum_{i=1}^{n} (c_i - c_{ri}) a_i \right|_K$$

$$\leqslant \sum_{i=1}^{n} |c_i - c_{ri}| \, |a_i|_K < nM \frac{\varepsilon}{nM} = \varepsilon.$$

Hence $\lim_{r \to \infty} b_r$ exists in K and is equal to $\sum_{i=1}^{n} c_i a_i$.

Now suppose that in the expressions (1), $c_{ri} = 0$ for all r whenever $i > 1$, that is, $b_r = c_{r1} a_1$ for all r. Given $\varepsilon > 0$ there is an $r_0 = r_0(\varepsilon)$ such that $|b_r - b_s|_K < \varepsilon |a_1|_K$ whenever $r, s > r_0$. Hence $|c_{r1} - c_{s1}| = |b_r - b_s|_K / |a_1|_K < \varepsilon$ whenever $r, s > r_0$ and so $\{c_{r1}\}$ is a Cauchy sequence. Assume that we have shown that if each term in a Cauchy sequence of elements of K can be written as a linear combination of $a_1, \ldots, a_{t-1}$, then for each i the coefficients of a_i form a Cauchy sequence of elements of k. Then assume that for each r,

$$b_r = c_{r1} a_1 + \cdots + c_{rt} a_t.$$

If $\{c_{rt}\}$ is a Cauchy sequence of elements of k then $\{b_r - c_{rt} a_t\}$ is a Cauchy sequence of elements of K. For, given $\varepsilon > 0$, there is an $r_0 = r_0(\varepsilon)$ such that $|b_r - b_s|_K < \varepsilon/2$ and $|c_{rt} - c_{st}| < \varepsilon/2 |a_t|_K$ whenever $r, s > r_0$: hence whenever $r, s > r_0$ we have

$$|(b_r - c_{rt} a_t) - (b_s - c_{st} a_t)|_K$$
$$\leqslant |b_r - b_s|_K + |c_{rt} - c_{st}| \, |a|_K < \varepsilon.$$

But, for each r,

$$b_r - c_{rt} a_t = c_{r1} a_1 + \cdots + c_{r,t-1} a_{t-1},$$

and by our assumption, $\{c_{ri}\}$ is a Cauchy sequence of elements of k for $i = 1, \ldots, t - 1$.

Assume, on the other hand, that $\{c_{rt}\}$ is not a Cauchy sequence. Then there exists an $\eta > 0$ and two sequences of integers $r_1 < r_2 < \cdots$ and $s_1 < s_2 < \cdots$ such that $|c_{r_i t} - c_{s_i t}| \geqslant \eta$ for all $i \geqslant 1$. Set

$$d_i = (c_{r_i t} - c_{s_i t})^{-1} (b_{r_i} - b_{s_i}).$$

Given $\varepsilon > 0$ there is an integer $i_0 = i_0(\varepsilon)$ such that $|b_{r_i} - b_{s_i}|_K < \varepsilon\eta$ whenever $i > i_0$. Then $|d_i|_K < \varepsilon\eta/\eta = \varepsilon$ whenever $i > i_0$ and $\lim_{i \to \infty} d_i = 0$. Thus the sequence $\{d_i - a_t\}$ converges and therefore is a Cauchy sequence. But

$$d_i - a_t = \sum_{j=1}^{t-1} \frac{c_{rij} - c_{sij}}{c_{r_i t} - c_{s_i t}} a_j.$$

By our assumption the coefficients of each a_j in these sums form a Cauchy sequence of elements of k and for each $j = 1, \ldots, t - 1$ there is an element $c_j \in k$ such that

$$\lim_{i \to \infty} \frac{c_{rij} - c_{sij}}{c_{rit} - c_{sit}} = c_j$$

Then $c_1 a_1 + \cdots + c_{t-1} a_{t-1} + a_t = 0$ which contradicts the linear independence of $a_1, \ldots, a_n$ over k. ‖

THEOREM 11. *Let k be a complete non-Archimedean valued field with valuation $| \ |$. Let K be a finite extension of k. Then there is unique prolongation of $| \ |$ to a valuation on K. In fact, if $| \ |_K$ is a prolongation of $| \ |$ to a valuation on K then for each $a \in K$,*

$$|a|_K = \sqrt[n]{|N_{K/k}(a)|}, \tag{2}$$

where $n = [K:k]$.

Proof. By Theorem 9 there is a prolongation of $| \ |$ to a valuation $| \ |_K$ on K. We shall show that it must be given by the formula (2). Let a be an element of K such that $|a|_K < 1$. Let $c_1, \ldots, c_n$ be a basis of K over k. For each positive integer r we have

$$a^r = a_{r1} c_1 + \cdots + a_{rn} c_n,$$

where $a_{ri} \in k$ for $i = 1, \ldots, n$ and $r \geqslant 1$. Since $|a^r|_K = |a|_K^r \to 0$ as $r \to \infty$ it follows from the proof of Theorem 10 that $\lim_{r \to \infty} a_{ri} = 0$ for $i = 1, \ldots, n$. Let F be a finite normal extension of k such that $k \subseteq K \subseteq F$, let $n_0 = [K:k]_s$, and let $\sigma_1, \ldots, \sigma_{n_0}$ be the distinct k-isomorphisms of K into F. Then

$$N_{K/k}(a^r) = \left(\prod_{i=1}^{n_0} \sigma_i(a^r) \right)^{[K:k]_i}$$

$$= \left(\prod_{i=1}^{n_0} (a_{r1}\sigma_i(c_1) + \cdots + a_{rn}\sigma_i(c_n)) \right)^{[K:k]_i}$$

$$= f(a_{r1}, \ldots, a_{rn}),$$

where $f(x_1, \ldots, x_n)$ is a homogeneous form in n variables whose coefficients are in K and do not depend on r. Hence $\lim_{r \to \infty} N_{K/k}(a)^r = \lim_{r \to \infty} N_{K/k}(a^r) = f(0, \ldots, 0) = 0$, which implies that $|N_{K/k}(a)| < 1$. Similarly, if $|a|_K > 1$ then $|N_{K/k}(a)| > 1$. Therefore, if $|N_{K/k}(a)| = 1$ we must have $|a|_K = 1$.

Now let a be an arbitrary element of K. If $a = 0$ then $|a|_K$ is certainly given by formula (2). Suppose $a \neq 0$. If we set $b = a^n/N_{K/k}(a)$ then $N_{K/k}(b) = N_{K/k}(a^n)/N_{K/k}(N_{K/k}(a)) = N_{K/k}(a)^n/N_{K/k}(a)^n = 1$. Hence $|b|_K = |a|_K^n/|N_{K/k}(a)| = 1$, that is, $|a|_K^n = |N_{K/k}(a)|$, from which we obtain formula (2). ‖

COROLLARY. *Let k be a complete non-Archimedean valuated field with valuation $|\ |$. Let C be an algebraic closure of k. Then $|\ |$ has a unique prolongation to a valuation on C.*

Proof. The existence of at least one prolongation follows from Theorem 9. Suppose that $|\ |_1$ and $|\ |_2$ are distinct prolongations of $|\ |$ to valuations on C. Then there is an element $a \in C$ such that $|a|_1 \neq |a|_2$ and so $|\ |_1$ and $|\ |_2$ are distinct prolongations of $|\ |$ to valuations on the finite extension $k(a)$ of k, contrary to Theorem 11. ‖

Consider now a complete valuated field k whose valuation $|\ |$ is Archimedean. Then char $k = 0$ and therefore k contains a subfield isomorphic to the field Q of rational numbers. For our purposes we may assume that this subfield is Q itself. The valuation $|\ |$ on k induces an Archimedean valuation $|\ |_Q$ on Q. By Theorem 2, $|\ |_Q$ is equivalent to ordinary absolute value. Since k is complete, it contains a subfield isomorphic to the completion of Q with respect to $|\ |_Q$. Hence we may assume that it contains as a subfield the field R of real numbers. If $R \neq k$ then k contains an element a such that $a^2 = -1$ (Exercise 22). In this case k has a subfield isomorphic to the field of complex numbers, and $|\ |$ induces on this subfield a valuation equivalent to ordinary absolute value. The fact that k is actually equal to this subfield is a consequence of the following Theorem 12. Its proof is given in a series of exercises at the end of this chapter.

THEOREM 12. *Let k be a complete Archimedean valuated field with valuation $|\ |$. Then, after replacing $|\ |$ by an equivalent valuation if necessary, there is an analytic isomorphism from k onto either the field of real numbers or the field of complex numbers valuated by ordinary absolute value.*

Roughly speaking, this theorem says that the only complete Archimedean valuated fields are the fields of real and complex numbers with ordinary absolute value as the valuation.

6. Relatively complete fields

Let k be a non-Archimedean valuated field. We assume that the valuation $|\ |$ on k is nontrivial. Let $\mathfrak{o}$ be the ring of integers of k and $\mathfrak{p}$ the ideal of nonunits of $\mathfrak{o}$. Let $\bar{k} = \mathfrak{o}/\mathfrak{p}$. If $a \in \mathfrak{o}$, we denote its residue class in $\bar{k}$ by $\bar{a}$. If $f(x) \in \mathfrak{o}[x]$ and if we replace each coefficient of $f(x)$ by its residue class in $\bar{k}$ we denote the resulting polynomial in $\bar{k}[x]$ by $\bar{f}(x)$.

Consider the following result, which may or may not hold for our valuated field k.

HENSEL'S LEMMA. *Let $f(x) \in \mathfrak{o}[x]$ and suppose that $\bar{f}(x) = G(x)H(x)$ where $G(x)$ and $H(x)$ are nonzero relatively prime polynomials in $\bar{k}[x]$. Then $f(x) = g(x)h(x)$ where $g(x), h(x) \in k[x]$ and $\deg g(x) = \deg G(x)$.*

DEFINITION. The valuated field k is said to be *relatively complete* if Hensel's lemma holds for k.

As we shall see, if k is complete then k is relatively complete. However, there are relatively complete fields which are not complete and we shall close this section with an example of such a valuated field. Also, as we shall see, there are non-Archimedean valuated fields which are not relatively complete.

THEOREM 13. (The Reducibility Criterion.) *Suppose that k is relatively complete and let $f(x) = a_0 + a_1 x + \cdots + a_n x^n \in k[x]$ with $a_n \neq 0$. Suppose that for some i with $0 < i < n$ we have $|a_n| < |a_i| = \max_{0 \leqslant j \leqslant n} |a_j|$. Then $f(x)$ is reducible in $k[x]$.*

Proof. Let i be the largest subscript such that $|a_i| = \max_{0 \leqslant j \leqslant n} |a_j|$. Then $i \neq n$ and $a_i \neq 0$. After dividing $f(x)$ by a_i, we may assume that $f(x) \in \mathfrak{o}[x]$ and that $a_i = 1$. If $G(x) = \bar{a}_0 + \bar{a}_1 x + \cdots + \bar{a}_{i-1} x^{i-1} + x^i$ and $H(x) = 1$, then $\bar{f}(x) = G(x)H(x)$ and $G(x)$ and $H(x)$ are relatively prime. Therefore, by Hensel's lemma, $f(x)$ has a factor in $k[x]$ of degree i and $0 < i < n$. ‖

THEOREM 14. *Let k be a relatively complete non-Archimedean valuated field with valuation $|\ |$, and let C be an algebraic closure of k. Then $|\ |$ has one and only one prolongation to a valuation on C.*

The existence of one prolongation follows from Theorem 9. However, we need more precise information than this simple fact of existence in order to show the uniqueness of the prolongation. Let K be a finite extension of k contained in C. Let $[K: k] = n$ and define the real-valued function $|\ |_K$ on K by

$$|a|_K = \sqrt[n]{|N_{K/k}(a)|}.$$

We shall show that $|\ |_K$ is a valuation on K: if it is, then it is certainly a prolongation of $|\ |$. We have $|0|_K = 0$ and $|a|_K > 0$ if $a \neq 0$ since $N_{K/k}(a) \neq 0$. Also, for $a, b \in K$,

$$|ab|_K = \sqrt[n]{|N_{K/k}(ab)|} = \sqrt[n]{|N_{K/k}(a)N_{K/k}(b)|}$$
$$= \sqrt[n]{|N_{K/k}(a)|}\ \sqrt[n]{|N_{K/k}(b)|} = |a|_K |b|_K.$$

It remains to show that $|a + b|_K \leqslant \max(|a|_K, |b|_K)$ for all $a, b \in K$, and to do this it is sufficient to show that if $a \in K$ and $|a|_K \leqslant 1$ then $|a + 1|_K \leqslant 1$ (see Exercise 4).

Let $f(x) = c_0 + c_1 x + \cdots + c_{m-1} x^{m-1} + x^m = \text{Irr}\,(k,a)$. By Theorem 26 of Chapter 1 there is an integer s such that $N_{K/k}(a) = (\pm c_0)^s$ and $|a|_K \leqslant 1$ implies that $|c_0| \leqslant 1$. Since $f(x)$ is irreducible in $k[x]$, by Theorem 13, we must have $|c_i| \leqslant 1$ for $0 < i < m$. Now $f(x - 1) = \text{Irr}\,(k, a + 1)$ and the constant term of this polynomial is $\pm\left(1 + \sum_{i=0}^{m-1}(\pm c_i)\right)$. For some integer t we have

$$N_{K/k}(a + 1) = \left[\pm\left(1 + \sum_{i=0}^{m-1}(\pm c_i)\right)\right]^t$$

and so

$$|a + 1|_K{}^n = |N_{K/k}(a + 1)| \leqslant \max(1, |c_0|, \ldots, |c_{m-1}|)^t \leqslant 1,$$

so that $|a + 1|_K \leqslant 1$.

In the proof of the uniqueness of the prolongation we shall need the

LEMMA. *Let k be as in Theorem* 14. *If* $f(x) = c_0 + c_1 x + \cdots + c_{n-1}x^{n-1} + x^n$ *is an irreducible polynomial in* $k[x]$, *then* $|c_j| \leqslant |c_0|^{(n-j)/n}$ *for* $j = 0, \ldots, n - 1$.

Proof. Let K be the splitting field of $f(x)$ over k in C. Let $b_1, \ldots, b_n$ be the roots of $f(x)$ in K, each listed as many times as its multiplicity. Let $|\ |_K$ be the valuation on K given above. Then for $i = 1, \ldots, n$ we have

$$|b_i|_K^{[K:k]} = |N_{K/k}(b_i)| = |N_{k(b_i)/k}(N_{K/k(b_i)}(b_i))|,$$
$$= |N_{k(b_i)/k}(b_i)|^{[K:k(b_i)]} = |c_0|^{[K:k(b_i)]}.$$

Since $[K:k]/[K:k(b_i)] = [k(b_i):k] = n$ we have, therefore, $|b_i|_K = |c_0|^{1/n}$. Now, except for sign, c_j is the $(n-j)$th elementary symmetric function of $b_1, \ldots, b_n$. Hence $|c_j| \leqslant |b_i|_K^{n-j} = |c_0|^{(n-j)/n}$. ‖

Proof of the uniqueness part of Theorem 14. Suppose that the valuation $|\ |$ on k has two distinct prolongations to valuations on C. Then there is a finite extension K of k in C such that $|\ |$ has two distinct prolongations to valuations on K. Let $[K:k] = n$. Then one prolongation of $|\ |$ is given by

$$|a|_K = \sqrt[n]{|N_{K/k}(a)|}, \qquad a \in K.$$

Let $|\ |_1$ be a valuation on K which prolongs $|\ |$ and is different from $|\ |_K$. Let $b \in K$ be such that $|b|_1 \neq |b|_K$. We may assume that $K = k(b)$. Let Irr $(k,b) = c_0 + c_1 x + \cdots + c_{n-1}x^{n-1} + x^n$: then $|b|_K = |c_0|^{1/n}$. Suppose that $|b|_1 < |b|_K$. Then for $j > 0$, $|c_j b^j|_1 = |c_j|\ |b|_1{}^j < |c_0|^{(n-j)/n}|b|_K{}^j = |c_0|^{(n-j)/n}|c_0|^{j/n} = |c_0|$ so that $0 = |0|_1 = |c_0 + c_1 b + \cdots + c_{n-1}b^{n-1} + b^n|_1 = |c_0| \neq 0$, a contradiction. Thus we must have $|b|_1 > |b|_K$. Then for $j \leqslant n - 1$,

$$|c_j b^j|_1 = |c_j|\ |b|_1{}^j = |c_j|\ |b|_1{}^n|b|_1{}^{j-n} < |c_j|\ |b|_1{}^n|b|_K^{j-n}$$
$$= |c_j|\ |c_0|^{(j-n)/n}|b|_1{}^n \leqslant |b^n|_1$$

so that

$$0 = |0|_1 = |c_0 + c_1 b + \cdots + c_{n-1}b^{n-1} + b^n|_1 = |b^n|_1 \neq 0,$$

a contradiction. Thus our assumption that | | has two distinct prolongations to valuations on C always leads to a contradiction. ‖

Looking back over the proof of Theorem 14 we see that Hensel's lemma was used only once, and then in an indirect way. We used the reducibility criterion of Theorem 13 to show that | | has a prolongation of a certain type to a valuation on a finite extension of k. In fact, we applied Theorem 13 only to a monic polynomial. Now, to prove Theorem 13 for monic polynomials we do not need the full force of Hensel's lemma. Theorem 13 for monic polynomials follows immediately if the following result holds for k.

WEAK HENSEL'S LEMMA. *Let $f(x)$ be a monic polynomial in $\mathfrak{o}[x]$ and assume that $\bar{f}(x) = G(x)H(x)$ where $G(x)$ and $H(x)$ are relatively prime polynomials in $\bar{k}[x]$ and $0 < \deg G(x) < \deg f(x)$. Then $f(x)$ is reducible in $k[x]$.*

If the Weak Hensel's lemma holds for k we call k *weakly relatively complete*. Since Theorem 13, for monic polynomials, holds for weakly relatively complete fields, we have already proved half of

THEOREM 15. *Let k be a non-Archimedean valued field with valuation | | and let C be an algebraic closure of k. Then | | has one and only one prolongation to a valuation on C if and only if k is weakly relatively complete.*

Proof. It remains to be shown that if | | has a unique prolongation to a valuation | $|_C$ on C then k is weakly relatively complete. Let $\sigma \in G(C/k)$ and define the real-valued function | $|_1$ on C by $|a|_1 = |\sigma(a)|_C$. Then | $|_1$ is a valuation on C: in particular, for all $a, b \in C$, $|a + b|_1 = |\sigma(a + b)|_C = |\sigma(a) + \sigma(b)|_C \leqslant \max(|\sigma(a)|_C, |\sigma(b)|_C) = \max(|a|_1, |b|_1)$. Furthermore, | $|_1$ is a prolongation of | |, and we can conclude that $|\sigma(a)|_C = |a|_C$ for all $a \in C$ and all $\sigma \in G(C/k)$.

Let $f(x)$ be a monic polynomial in $\mathfrak{o}[x]$ and suppose that $\bar{f}(x) = G(x)H(x)$ where $G(x)$ and $H(x)$ are relatively prime polynomials in $\bar{k}[x]$ and $0 < \deg G(x) < \deg f(x)$. Then, since $\deg \bar{f}(x) = \deg f(x)$ we have $0 < \deg H(x) < \deg f(x)$. Assume that $f(x)$ is irreducible in $k[x]$ and factor $f(x)$ in $C[x]$:
$$f(x) = (x - a_1) \cdots (x - a_n).$$

For each i and j there is a k-isomorphism σ from $k(a_i)$ onto $k(a_j)$ such that $\sigma(a_i) = a_j$. Since C is a normal extension of k, and since the proofs of Propositions 1 and 2 of Section 10 of Chapter 2 require only the normality of the extensions, there is a k-automorphism of C which coincides with σ on $k(a_1)$. Hence $|a_j|_C = |\sigma(a_i)|_C = |a_i|_C$. Therefore $|a_i|_C$ is the same for all i and since $f(x)$ has coefficients in $\mathfrak{o}$ this implies that $|a_i|_C \leqslant 1$ for $i = 1, \ldots, n$. Thus $\bar{f}(x) = (x - \bar{a}_1) \cdots (x - \bar{a}_n)$ in $\bar{C}[x]$, where $\bar{C}$ is the residue class field of $(C, | |_C)$. Since neither $G(x)$ or $H(x)$ is a constant there are integers i and j such that $x - \bar{a}_i$ divides $G(x)$ and $x - \bar{a}_j$ divides $H(x)$ in $\bar{C}[x]$.

Let $\mathfrak{O}$ be the ring of integers of $(C, |\ |_C)$ and let $\mathfrak{P}$ be the ideal of nonunits of $\mathfrak{O}$. Let $g(x) \in \mathfrak{o}[x]$ be such that $\bar{g}(x) = G(x)$ and let $g_1(x) \in \mathfrak{O}[x]$ be such that

$$g(x) - (x - a_i)g_1(x) \in \mathfrak{P}[x].$$

If we apply σ to the coefficients of these polynomials we obtain

$$g(x) - (x - a_j)(\sigma g_1)(x) \in \mathfrak{P}[x],$$

since σ maps elements of $\mathfrak{P}$ onto elements of $\mathfrak{P}$. Thus

$$G(x) - (x - \bar{a}_j)(\overline{\sigma g_1})(x) = 0.$$

Hence $x - \bar{a}_j$ divides both $G(x)$ and $H(x)$ in $\bar{C}[x]$, which contradicts the fact that $G(x)$ and $H(x)$ are relatively prime in $\bar{k}[x]$. Therefore $f(x)$ must be reducible in $k[x]$ and the Weak Hensel's lemma holds for k. ‖

We are still left with an unanswered question, namely, that of the relation between relative completeness and weak relative completeness. It is clear that if k is relatively complete then it is weakly relatively complete. It is surprising that the converse is also true, that is, the following theorem holds.

THEOREM 16. *A non-Archimedean valuated field is relatively complete if and only if it is weakly relatively complete.*

We need to prove that weak relative completeness implies relative completeness. In order to do this we must make a careful examination of $f(x)$ where $f(x) \in \mathfrak{o}[x]$. We say that $f(x) \in \mathfrak{o}[x]$ is a *primitive* polynomial if at least one of its coefficients is a unit in $\mathfrak{o}$.

LEMMA 1. *The product of primitive polynomials in $\mathfrak{o}[x]$ is primitive.*

Proof. It is enough to prove that the product of two primitive polynomials is primitive. Let

$$f(x) = a_0 + a_1 x + \cdots + a_n x^n, \qquad g(x) = b_0 + b_1 x + \cdots + b_m x^m$$

be primitive polynomials in $\mathfrak{o}[x]$. Assume that a_i is the first coefficient of $f(x)$ from the left that is a unit in $\mathfrak{o}$ and that b_j is the first coefficient of $g(x)$ from the left that is a unit in $\mathfrak{o}$. The coefficient of x^{i+j} in $f(x)g(x)$ is

$$(a_0 b_{i+j} + \cdots + a_{i-1} b_{j+1}) + a_i b_j + (a_{i+1} b_{j-1} + \cdots + a_{i+j} b_0).$$

This is a unit in $\mathfrak{o}$ since $a_i b_j$ is a unit and the expressions in parentheses are nonunits. Therefore $f(x)g(x)$ is primitive. ‖

LEMMA 2. *Let $f(x)$ be a primitive polynomial in $\mathfrak{o}[x]$ and suppose that $f(x) = g_1(x)h_1(x)$ where $g_1(x), h_1(x) \in k[x]$. Then there are primitive polynomials $g(x)$ and $h(x)$ in $\mathfrak{o}[x]$ such that $f(x) = g(x)h(x)$ and $g(x)$ and $h(x)$ are constant multiples of $g_1(x)$ and $h_1(x)$, respectively.*

Proof. There are nonzero constants $c, d \in k$ such that $g_1(x) = cg(x)$ and $h_1(x) = dh(x)$ where $g(x)$ and $h(x)$ are primitive polynomials in $\mathfrak{o}[x]$. Then $f(x) = cdg(x)h(x)$. By Lemma 1, $g(x)h(x)$ is primitive. Let $f(x) = a_0 + a_1 x + \cdots + a_n x^n$ and $g(x)h(x) = b_0 + b_1 x + \cdots + b_n x^n$. Let i and j be such that a_i and b_j are units in $\mathfrak{o}$. Then $a_i = cdb_i$ and $a_j = cdb_j$, and so

$$|cd| = |a_i|\,|b_i|^{-1} \geqslant 1, \qquad |cd| = |a_j|\,|b_j|^{-1} \leqslant 1.$$

Thus cd is a unit in $\mathfrak{o}$ and if we write $g(x)$ for $cdg(x)$ then $f(x) = g(x)h(x)$ and $g(x)$ and $h(x)$ are primitive. ‖

LEMMA 3. *Let $f(x)$ be a primitive polynomial in $\mathfrak{o}[x]$ which is irreducible in $k[x]$. If k is weakly relatively complete, $\bar{f}(x)$ is a constant multiple of a power of an irreducible polynomial in $\bar{k}[x]$.*

Proof. Let $f(x) = a_0 + a_1 x + \cdots + a_n x^n$. If C is an algebraic closure of k then $f(x)$ can be factored in $C[x]$ as

$$f(x) = a_n(x - b_1) \cdots (x - b_n).$$

Arguing as we have done previously, using the fact that the valuation $|\ |$ on k has a unique prolongation to a valuation $|\ |_C$ on C and our assumption that $f(x)$ is irreducible in $k[x]$, we conclude that $|b_i|_C$ is the same for all i. If both a_n and a_0 are nonunits in $\mathfrak{o}$ then $f(x)$ is not primitive. Hence we are left with three cases to consider.

Case 1. a_0 is a unit and a_n is a nonunit. We have $1 = |a_0| = |a_n|\,|b_i|_C^n < |b_i|_C^n$ for each i. If $1 \leqslant j \leqslant n - 1$ then $|a_n|\,|b_i|_C^{n-j} = |b_i|_C^{-j} < 1$. Hence any product of a_n and fewer than n of the b_i's has valuation less than one, that is, it is a nonunit in $\mathfrak{o}$. Therefore every coefficient of $f(x)$ is a nonunit except a_0 and so $\bar{f}(x) = \bar{a}_0$.

Case 2. a_0 is a nonunit and a_n is a unit. Since $|b_i|_C^n = |a_n|\,|b_i|_C^n = |a_0| < 1$, every b_i is a nonunit. Hence $\bar{f}(x) = \bar{a}_n x^n$.

Case 3. a_0 and a_n are both units. In this case we may divide $f(x)$ by a_n and continue to have a primitive polynomial in $\mathfrak{o}[x]$, and we can assume that $f(x)$ is monic. If $\bar{f}(x)$ is not a constant multiple of a power of an irreducible polynomial in $\bar{k}[x]$, then $\bar{f}(x) = G(x)H(x)$ where $G(x)$ and $H(x)$ are relatively prime nonconstant polynomials in $\bar{k}[x]$. But then, by the Weak Hensel's lemma, $f(x)$ is reducible in $k[x]$, which contradicts the hypothesis of the lemma. Therefore, in this case also, $\bar{f}(x)$ must have the required form. ‖

Completion of the proof of Theorem 16. Assume that the non-Archimedean valuated field k is weakly relatively complete. Let $f(x) \in \mathfrak{o}[x]$ and suppose that $\bar{f}(x) = G(x)H(x)$ where $G(x)$ and $H(x)$ are non-zero relatively prime polynomials in $\bar{k}[x]$. Then $f(x)$ is primitive. If we factor $f(x)$ into irreducible factors in $k[x]$ we can assume, by Lemma 2, that these factors are primitive polynomials in $o[x]$. Let $f(x) = p_1(x) \cdots p_n(x)$ where for each i, $p_i(x)$ is a primitive polynomial in $o[x]$ which is irreducible in $k[x]$. By Lemma 3, for each i, $\bar{p}_i(x)$ is a constant multiple of a power of an irreducible polynomial in $\bar{k}[x]$. Thus, since $G(x)$ and $H(x)$ are relatively prime in $\bar{k}[x]$, we can number the $p_i(x)$ so that $G(x) = \bar{p}_1(x) \cdots \bar{p}_t(x)$ and $H(x) = \bar{p}_{t+1}(x) \cdots \bar{p}_n(x)$. It also follows from the proof of Lemma 3 that for each i, $\bar{p}_i(x)$ is a constant or $\deg \bar{p}_i(x) = \deg p_i(x)$.

Now, we can number the $p_i(x)$ so that $\deg \bar{p}_i(x) = \deg p_i(x)$ for $i = 1, \ldots, s$ and $\bar{p}_i(x)$ is a constant for $i = s + 1, \ldots, t$. Let $g(x) = dp_1(x) \cdots p_s(x)$ where d is chosen (in k) so that $\bar{g}(x) = G(x)$ and let $h(x) = d^{-1}p_{s+1}(x) \cdots p_n(x)$. Then $f(x) = g(x)h(x)$ and $\deg g(x) = \deg G(x)$. ‖

It follows from the corollary to Theorem 11 and from Theorem 14 that a complete non-Archimedean field is relatively complete. The converse of this statement is not true and we close this section with an example of a relatively complete valued field which is not complete.

Let k be the completion of the field of rational functions over $GF(p)$ with respect to the valuation $| \ |_x$ (see Section 2). Let C be an algebraic closure of $GF(p)$ and let F be the completion of the field of rational functions over C with respect to $| \ |_x$. We may assume that $k \subseteq F$. For each $n \geqslant 2$ let ζ_n be a primitive $(p^n - 1)$st root of unity in C and let K be the extension of k in F obtained by adjoining to k the set $\{\zeta_n \mid n \geqslant 2\}$. For each integer $r \geqslant 2$ let $a_r = \zeta_2 x^2 + \zeta_3 x^3 + \cdots + \zeta_r x^r$. Then $|a_{r+1} - a_r|_x = d^{r+1} \to 0$ as $r \to \infty$ and so $\{a_r\}$ is a Cauchy sequence of elements of K. The limit of this sequence in F is the power series $\sum_{n=2}^{\infty} \zeta_n x^n$. The representation of an element of F as a power series in x is unique. Any given element of K belongs to a subfield of K that is obtained by adjoining a finite number of ζ_n to k and so in the representation of such an element as a power series in x only a finite number of the ζ_n can appear in its coefficients. Therefore $\sum_{n=2}^{\infty} \zeta_n x^n \notin K$. Thus K is not complete. However, K is an algebraic extension of a complete non-Archimedean valued field and therefore K is relatively complete.

7. Prolongations of valuations, continued

Let k be a valued field with valuation $| \ |$. Let K be a finite extension of the field k. We know that $| \ |$ can be prolonged to a valuation on K: this follows from Theorem 9 in the non-Archimedean case and from Theorem 12 in the Archimedean case. Now we shall be concerned with counting the number of prolongations. We know from Theorem 11 in Chapter 1 that K can be obtained from k by a separable extension followed by a purely inseparable extension. Thus we can do our counting in two stages. We will deal with separable extensions in this section and leave the purely inseparable case to the reader (see Exercise 33).

THEOREM 17. *Let K be a finite separable extension of a field k and let $|\ |$ be a valuation on k. Let a be a primitive element of K with respect to k and let $f(x) = \text{Irr}(k,a)$. Let $\hat{k}$ be the completion of k with respect to $|\ |$ and assume that $f(x)$ factors into the product of r distinct nonconstant irreducible factors in $\hat{k}[x]$. Then $|\ |$ has exactly r distinct prolongations to valuations on k.*

Proof. We shall show first that to every prolongation $|\ |_1$ of $|\ |$ to a valuation on K there corresponds an irreducible factor of $f(x)$ in $\hat{k}[x]$. Let $\hat{K}_1$ be the completion of K with respect to $|\ |_1$ and let C be an algebraic closure of $\hat{K}_1$. The valuation $|\ |_1$ can be uniquely prolonged to a valuation on C. We can prolong $|\ |$ to a valuation on $\hat{k}(a)$ and this must coincide with the restriction to $\hat{k}(a)$ of the valuation on C. We have $K \subseteq k(a) \subseteq \hat{K}_1$ and since $\hat{k}(a)$ is complete we conclude that $\hat{K}_1 = \hat{k}(a)$. Thus $|\ |_1$ determines an irreducible factor of $f(x)$ in $\hat{k}[x]$, namely, $\text{Irr}(\hat{k},a)$. Note that C is an algebraic closure of $\hat{k}$ and that the valuation on C is precisely the unique prolongation of the valuation on $\hat{k}$ to a valuation on C.

It should be fairly clear at this stage how the prolongations of $|\ |$ to valuations on K are obtained. Let C be an algebraic closure of $\hat{k}$ and prolong $|\ |$ uniquely to a valuation on C. Let $p(x)$ be one of the irreducible factors of $f(x)$ in $\hat{k}[x]$ and let $\hat{a}$ be a root of $p(x)$ in C. Then $\hat{k}(\hat{a})$ has a unique valuation prolonging $|\ |$. In addition, $\hat{k}(\hat{a})$ contains a k-isomorphic image of K, say $\sigma(K)$: this is precisely $k(\hat{a})$ and we have $\sigma(a) = \hat{a}$. The valuation on $\hat{k}(\hat{a})$ induces a valuation $|\ |'$ on $\sigma(K)$. If $b \in K$ we set $|b| = |\sigma(b)|'$: then $|\ |$ is a valuation on K which prolongs the valuation on k.

Let $p(x)$ and $q(x)$ be different irreducible factors of $f(x)$ in $\hat{k}[x]$. Each of these factors determines a valuation on K according to the preceding paragraph. We shall show that these valuations must be distinct. Since they both prolong the valuation on k, they cannot be equivalent unless they coincide. Let $\hat{a}$ be a root of $p(x)$ in C and $\hat{b}$ a root of $q(x)$ in C. Let $\sigma(K) = k(\hat{a})$ with $\sigma(a) = \hat{a}$ and $\tau(K) = k(\hat{b})$ with $\tau(a) = \hat{b}$. Let $|\ |$ be the prolongation of the valuation on k to a valuation on C. Let $p(x) = c_0 + c_1 x + \cdots + c_s x^s$ and for $i = 0, \ldots, s$ let $\{c_{in}\}$ be a Cauchy sequence of elements of k having c_i as a limit. Set $p_n(x) = c_{on} + c_{1n} x + \cdots + c_{sn} x^s$. Then $\{p_n(\hat{a})\}$ is a Cauchy sequence of elements of $\sigma(K)$ and $\{p_n(\hat{b})\}$ is a Cauchy sequence of elements of $\tau(K)$. For each n, $\sigma(p_n(a)) = p_n(\hat{a})$ and $\tau(p_n(a)) = p_n(\hat{b})$. Let $|\ |_1$ be the valuation on K determined by $p(x)$ and $|\ |_2$ that determined by $q(x)$. Then $\lim_{n \to \infty} |p_n(a)|_1 = \lim_{n \to \infty} |p_n(\hat{a})| = |\lim_{n \to \infty} p_n(\hat{a})| = |p(\hat{a})| = |0| = 0$. Since $p(x)$ and $q(x)$ are distinct and irreducible in $\hat{k}[x]$ and $q(\hat{b}) = 0$ we have $p(\hat{b}) \neq 0$. Then $\lim_{n \to \infty} |p_n(a)|_2 = |p(\hat{b})| \neq 0$. Thus $|\ |_1$ and $|\ |_2$ are distinct valuations on K. Note that we may have $k(\hat{a}) = k(\hat{b})$ (see the example after the proof).

Finally we must show that if $\hat{a}$ and $\hat{b}$ are roots of the same irreducible factor of $f(x)$ in $\hat{k}[x]$ then they determine the same valuation on K. Let $\sigma(K) = k(\hat{a})$ with $\sigma(a) = \hat{a}$ and $\tau(K) = k(\hat{b})$ with $\tau(a) = \hat{b}$. There is a $\hat{k}$-isomorphism ρ from $\hat{k}(\hat{a})$ onto $\hat{k}(\hat{b})$ such that $\rho(\hat{a}) = \hat{b}$ and ρ is analytic since $|\ |$ has a unique prolongation to a valuation on $\hat{k}(\hat{b})$. If $g(x) \in \hat{k}[x]$ then $\rho(g(\hat{a})) = g(\hat{b})$. If $c \in K = k(a)$ then $c = r(a)$ where $r(x) \in k[x]$. We have $\rho\sigma(c) = \rho\sigma(r(a)) = \rho(r(\hat{a})) = r(\hat{b}) = \tau(r(a)) = \tau(c)$. Thus, if $|\ |_1$ and $|\ |_2$ are the valuations on K obtained as above, using $\hat{a}$ and $\hat{b}$, respectively, we have $|c|_1 = |\sigma(c)| = |\rho\sigma(c)| = |\tau(c)| = |c|_2$ so that $|\ |_1$ and $|\ |_2$ are the same valuations on K. ‖

As an example, consider the field Q of rational numbers and the problem of prolonging certain valuations on Q to valuations on $Q(\sqrt{2})$. Let $|\ |$ be the ordinary absolute value on Q. The completion of Q with respect to $|\ |$ is the field R of real numbers. In $R[x]$, $f(x) = \text{Irr}(Q, \sqrt{2}) = x^2 - 2$ factors into $(x - \sqrt{2})(x + \sqrt{2})$. Hence there are two prolongations of $|\ |$ to valuations on $Q(\sqrt{2})$. There are two Q-isomorphisms from $Q(\sqrt{2})$ into R, one taking $\sqrt{2}$ onto $\sqrt{2}$ and the other taking $\sqrt{2}$ onto $-\sqrt{2}$. Therefore the two prolongations of $|\ |$ are given by $|a + b\sqrt{2}|' = |a + b\sqrt{2}|$ and $|a + b\sqrt{2}|'' = |a - b\sqrt{2}|$.

Now consider the 2-adic valuation $|\ |_2$ on Q. The completion of Q with respect to $|\ |_2$ is the field Q_2 of 2-adic numbers. The integer 2 is a prime in the ring of integers of Q_2 and therefore $x^2 - 2$ is irreducible in $Q_2[x]$: this follows from the Eisenstein irreducibility criterion (a general form of this criterion will be given as Theorem 6 in Chapter 4). Hence $|\ |_2$ has only one prolongation to a valuation $|\ |_2'$ on $Q(\sqrt{2})$. We note, however, that there are two Q-isomorphisms of $Q(\sqrt{2})$ into $Q_2(\sqrt{2})$, σ which maps $\sqrt{2}$ onto $\sqrt{2}$ and τ which maps $\sqrt{2}$ onto $-\sqrt{2}$. If $a + b\sqrt{2} \in Q(\sqrt{2})$ then

$$|\sigma(a + b\sqrt{2})|_2 = |a + b\sqrt{2}|_2$$

$$= \sqrt{|N_{Q_2(\sqrt{2})/Q_2}(a + b\sqrt{2})|_2}$$

$$= \sqrt{|a^2 - 2b^2|_2}$$

while

$$|\tau(a + b\sqrt{2})|_2 = |a - b\sqrt{2}|_2$$

$$= \sqrt{|N_{Q_2(\sqrt{2})/Q_2}(a - b\sqrt{2})|_2}$$

$$= \sqrt{|a^2 - 2b^2|_2}.$$

Thus, using either of these Q-isomorphisms, we arrive at the unique prolongation of $|\ |_2$ to a valuation on $Q(\sqrt{2})$ given by

$$|a + b\sqrt{2}|_2' = \sqrt{|a^2 - 2b^2|_2} \ .$$

For the remainder of this section we continue under the hypothesis of Theorem 17. Let $|\ |_1, \ldots, |\ |_r$ be the prolongations of $|\ |$ to distinct valuations on K. For $i = 1, \ldots, r$ let $\hat{K}_i$ be the completion of K with respect to $|\ |_i$ and let $d_i(K/k) = [\hat{K}_i : \hat{k}]$. Then, as an immediate consequence of the proof of Theorem 17, we have

$$[K : k] = \sum_{i=1}^{r} d_i(K/k).$$

Finally, we shall prove one theorem which illustrates the relation between the "local" and "global" aspects of a valuated field.

THEOREM 18. *For all $a \in K$,*

$$N_{K/k}(a) = \prod_{i=1}^{r} N_{\hat{K}_i/\hat{k}}(a)$$

and

$$T_{K/k}(a) = \sum_{i=1}^{r} T_{\hat{K}_i/\hat{k}}(a)$$

Proof. Suppose that $K = k(a)$ and let $f(x) = \text{Irr}\ (k,a)$. If we compare the coefficients of $f(x)$ with the coefficients of the product of the irreducible factors of $f(x)$ in $\hat{k}[x]$ we will obtain the formulas of the theorem for this particular element of K. Now let b be an arbitrary element of K. If $b = 0$ the formulas of the theorem hold, and we assume that $b \neq 0$. Consider the fields $k(b(a + c))$ where c runs through k. Since $|\ |$ is a nontrivial valuation, k must have infinitely many elements. Since K/k is separable there are only a finite number of fields between k and K (see the proof of Theorem 24 in Chapter 1). Hence for some c and d in k with $c \neq d$ we have $k(b(a + c)) = k(b(a + d)) = L$. Then $b(c - d) \in L$ and since $c - d \in k$ and $c - d \neq 0$ we have $b \in L$. Thus $ba \in L$ and since $b \neq 0$ we have $a \in L$ and so $L = k(a) = K$. Therefore $K = k(b(a + c))$. We also have $K = k(a + c)$ since $c \in k$ and so

$$N_{K/k}(b) = \frac{N_{K/k}(b(a + c))}{N_{K/k}(a + c)} = \frac{\displaystyle\prod_{i=1}^{r} N_{\hat{K}_i/\hat{k}}(b(a + c))}{\displaystyle\prod_{i=1}^{r} N_{\hat{K}_i/\hat{k}}(a + c)}$$

$$= \prod_{i=1}^{r} \frac{N_{\hat{K}_i/\hat{k}}(b(a + c))}{N_{\hat{K}_i/\hat{k}}(a + c)} = \prod_{i=1}^{r} N_{\hat{K}_i/\hat{k}}(b),$$

and the formula for the trace is obtained in a similar manner. ‖

EXERCISES

Section 1

1. Let $| \ |$ be a valuation on a field k. Show that
   (a) $|1| = |-1| = 1$,
   (b) $|-a| = |a|$,
   (c) if $a \neq 0$, then $|a^{-1}| = |a|^{-1}$,
   (d) $|a - b| \leqslant |a| + |b|$,
   (e) $| \ |a| - |b| \ | \leqslant |a - b|$.

2. Let $| \ |$ be a non-Archimedean valuation on a field k. Show that if $|a| \neq |b|$ then $|a + b| = \max(|a|, |b|)$. Show that if $a_1, \ldots, a_n \in k$ and if $|a_1 + \cdots + a_n| < |a_i|$ for $i = 1, \ldots, n$, then for some i and j with $i \neq j$ we have $|a_i| = |a_j|$.

3. Show that the only valuation on a finite field is the trivial valuation.

4. Let $| \ |$ be a real-valued function defined on a field k such that $|a| \geqslant 0$ for all $a \in k$, $|a| = 0$ if and only if $a = 0$, and $|ab| = |a| \, |b|$ for all $a, b \in k$. Show that $| \ |$ is a non-Archimedean valuation on k if and only if $|a| \leqslant 1$ implies that $|1 + a| \leqslant 1$.

5. Let D be an integral domain, k a field of quotients of D, and $\mathfrak{p}$ a principal prime ideal of D such that $\bigcap_{n=1}^{\infty} \mathfrak{p}^n = 0$. Let c be a real number such that $0 < c < 1$. For $a \in D$, $a \neq 0$, set $|a| = c^n$ where n is determined by the fact that $a \in \mathfrak{p}^n$ but $a \notin \mathfrak{p}^{n+1}$. Set $|0| = 0$ and for $a, b \in D$, $b \neq 0$, set $|a/b| = |a| \, |b|^{-1}$. Show that $| \ |$ is a non-Archimedean valuation on k.

6. Show that an Archimedean valuation and a non-Archimedean valuation on the same field cannot be equivalent.

7. Let k be a field with a non-Archimedean valuation $| \ |$. A triple (a,b,c) of elements of k is called a *triangle* and the pairs (a,b), (b,c), and (c,a) its *sides*. The *length* of the side (a,b) is defined to be $|a - b|$. A triangle is called *isosceles* if two of its sides have the same length: the third side is then called its *base*. If $a \in k$ and if r is a real number then the set $\{x \in k \mid |x - a| < r\}$ is called the *open circle* with *center* a and *radius* r. If the strong inequality is replaced by weak inequality then we obtain the *closed circle* with center a and radius r. Two circles with the same center are said to be *concentric*. Verify the following statements:
   (a) Each triangle of k is isosceles and the length of its base is not greater than the length of its other two sides.
   (b) Each element of a circle is its center.
   (c) Two circles which have an element in common are concentric.
   (d) Two open (closed) circles are either disjoint or identical.
   (e) If an element a of k does not belong to a circle then $|x - a|$ is the same for all x in the circle.

(f) The circumference of a closed circle of center a and radius r, that is, the set $\{x \in k \mid |x - a| = r\}$, is a disjoint union of open circles of radius r.

(g) All circles, and the circumferences of closed circles, are both open and closed sets in the topology on k determined by $|\ |$.

8. By an *Artin valuation* on a field k we mean a real-valued function $|\ |$ defined on k which satisfies (i) and (ii) in the definition of valuation given in Section 1, and has the property that there is a constant $c \geqslant 1$ such that if $a \in k$ and $|a| \leqslant 1$ then $|1 + a| \leqslant c$. If we define equivalence of Artin valuations in the same way as we defined equivalence of ordinary valuations, then the result of Theorem 1 continues to hold. Verify the following statements:

(a) An ordinary valuation is an Artin valuation.

(b) Every Artin valuation is equivalent to an ordinary valuation.

(c) For an Artin valuation $|\ |$ we may take $c = \max(|1|, |2|)$.

Section 2

9. Let K be the field Q of rational numbers or the field $k(x)$ of rational functions over an arbitrary field k. Show that there is a set V of inequivalent valuations (other than the set consisting of the trivial valuation alone) on K such that

$$\prod_{|\ | \in V} |a| = 1 \qquad \text{for all nonzero } a \in K.$$

10. Let k be a non-Archimedean valuated field with valuation $|\ |$. Let c be a positive real number. If $f(x) \in k[x]$ and $f(x) = a_0 + a_1 x + \cdots + a_n x^n \neq 0$ set

$$|f(x)|^* = \max_{0 \leqslant i \leqslant n} |a_i| \, c^i.$$

Set $|0|^* = 0$, and if $f(x)$, $g(x) \in k[x]$ with $g(x) \neq 0$ set $|f(x)/g(x)|^* = |f(x)|^* |g(x)|^{*-1}$. Prove that $|\ |^*$ is a non-Archimedean valuation on $k(x)$ such that $|a|^* = |a|$ for all $a \in k$.

Section 3

11. Show that if a sequence of elements of a valuated field converges then it has a unique limit.

12. Prove, in detail, the proposition at the beginning of Section 3.

13. Let $|\ |$ be a non-Archimedean valuation on a field k. Show that a sequence $\{a_n\}$ of elements of k is a Cauchy sequence if and only if $\lim_{n \to \infty} (a_{n+1} - a_n) = 0$.

14. Let k be a valuated field and let $\{a_n\}$ and $\{b_n\}$ be convergent sequences of elements of k. Show that $\{a_n b_n\}$ and $\{a_n + b_n\}$ are convergent sequences of elements of k such that

$$\lim_{n \to \infty} (a_n + b_n) = \lim_{n \to \infty} a_n + \lim_{n \to \infty} b_n,$$

$$\lim_{n \to \infty} a_n b_n = \left(\lim_{n \to \infty} a_n \right) \left(\lim_{n \to \infty} b_n \right)$$

15. Let k be a complete non-Archimedean valuated field. If we define convergence of series of elements of k in the usual manner, show that $\Sigma\, a_n$ converges if and only if $\lim_{n \to \infty} a_n = 0$.

16. Complete the details of the proof of Theorem 4.

17. Let K and L be valuated fields with valuations $|\ |_K$ and $|\ |_L$, respectively, and let σ be an analytic isomorphism from K onto L. Show that σ can be extended to an analytic isomorphism from the completion of K onto the completion of L.

18. Develop a theory of infinite products in a complete valuated field.

Section 4

19. Determine the residue class field of $F(x)$ with respect to the valuation $|\ |_\infty$.

Section 5

20. Let k be a complete Archimedean valuated field with valuation $|\ |$. Let $a \in k$. Show that if $4\,|a| < |4|$ then $1 + a = b^2$ for some $b \in k$.

21. Let k be a complete Archimedean valuated field with valuation $|\ |$. Show that there is a prolongation of $|\ |$ to a valuation $|\ |_1$ on $k(i)$ where $i^2 = -1$. In fact, show that for $a \in k(i)$, $|a|_1 = \sqrt{|N_{k(i)/k}(a)|}$. (Note that this is not a consequence of Theorem 11, which was concerned with the non-Archimedean case.)

22. Let k be as in the discussion preceding Theorem 12. Show that if -1 is not a square in k then $k = R$. Therefore, if $k \neq R$ we may assume, in the discussion given there, that k contains the field C of complex numbers and that $|\ |$ induces ordinary absolute value on C.

23. Continuing from Exercise 22, show that if $k \neq C$ then there is an element $a \in k$, $a \notin C$, such that $|b - a| \geqslant |a| = t > 0$ for all $b \in C$.

24. Continuing from Exercise 23, show that if $k \neq C$ and if $a \in k$, $a \notin C$, meets the requirements of Exercise 23, then for all integers n we have $|n| \leqslant 2t$. Conclude from this that $|\ |$ is non-Archimedean, which is a contradiction. Note that we have now proved Theorem 12.

25. This exercise provides us with an alternate proof of Theorem 11. Let k be a complete non-Archimedean valuated field with valuation $|\ |$. Let K be a finite extension of k. Let $|\ |_1$ and $|\ |_2$ be two prolongations of $|\ |$ to valuations on K.

(a) Show that a sequence of elements of K is a Cauchy sequence with respect to $|\ |_1$ if and only if it is a Cauchy sequence with respect to $|\ |_2$.

(b) Show that $|\ |_1$ and $|\ |_2$ must be equivalent and conclude that they coincide.

Section 6

26. Let k be a complete non-Archimedean valuated field with valuation $|\ |$. In this exercise we give an alternate proof of Hensel's lemma for k. In the statement of Hensel's lemma we may assume that one of $G(x)$ and $H(x)$, say $G(x)$, is monic. Let $g_0(x)$, $h_0(x) \in \mathfrak{o}[x]$ be such that $\bar{g}_0(x) = G(x)$ and $\bar{h}_0(x) = H(x)$, and $g_0(x)$ is monic. Show that we may assume that $\deg g_0(x) + \deg h_0(x) \leqslant \deg f(x)$.

(a) Show that there is an element $b \in \mathfrak{o}$ with $|b| < 1$ such that $f(x) - g_0(x)h_0(x) \in b\mathfrak{o}[x]$ and $g_0(x)A(x) + h_0(x)B(x) - 1 \in b\mathfrak{o}[x]$ for polynomials $A(x)$, $B(x) \in \mathfrak{o}[x]$.

(b) Show that there are sequences $\{g_n(x)\}$ and $\{h_n(x)\}$ of polynomials in $\mathfrak{o}[x]$ such that for each positive integer n,

$$f(x) - g_n(x)h_n(x) \in b^{n+1}\mathfrak{o}[x],$$
$$\deg g_n(x) + \deg h_n(x) \leqslant \deg f(x),$$
$$g_n(x) - g_{n-1}(x), \qquad h_n(x) - h_{n-1}(x) \in b^n\mathfrak{o}[x],$$

and $g_n(x)$ is monic.

(c) For a fixed i let a_{in} be the coefficient of x^i in $g_n(x)$. Show that $\{a_{in}\}$ is a Cauchy sequence of elements of k. Let $a_i = \lim_{n \to \infty} a_{in}$ and $g(x) = \Sigma\, a_i x^i$. Define $h(x)$ in a similar manner, and show that $g(x)$ and $h(x)$ satisfy the conclusion of Hensel's lemma.

27. Let a be a rational integer and suppose that a is a quadratic residue modulo the odd prime p, that is, suppose that there is an integer b such that $a \equiv b^2 \pmod{p}$. Show that a is a square in the field of p-adic numbers. (Hint. Use Hensel's lemma.)

28. Let $a \equiv 1 \pmod 8$, where a is a rational integer. Show that a is a square in the field of 2-adic numbers.

29. Show that a finite extension of a relatively complete non-Archimedean valuated field is relatively complete.

30. Show that an algebraically closed non-Archimedean valuated field is relatively complete.

31. Show by a simple example that not every non-Archimedean valuated field need be relatively complete.

32. Let k be a relatively complete field and let $K = k(a)$ and $L = k(b)$ where Irr $(k,a) = $ Irr (k,b). Show that there is an analytic k-isomorphism σ from K onto L such that $\sigma(a) = b$.

Section 7

33. Let k be a valuated field with valuation $|\ |$ and let K be a purely inseparable extension of k. Show that $|\ |$ has a unique prolongation to a valuation on K.

34. Let $K = Q(i)$ where $i^2 = -1$. Determine the prolongations of the p-adic valuation $|\ |_p$ on Q to valuations on K.

35. Do the same for $K = Q(\sqrt{-3})$ and $K = Q(\sqrt{-5})$.

CHAPTER 4

Extensions of Valuated Fields

1. Ramification and residue class degree

Let k be a non-Archimedean valuated field and let K be a finite extension of k. We shall denote the valuation on K by $|\ |$ and we shall use the following notation:

$n = [K:k]$,
$\mathfrak{o}$ = the ring of integers of k,
$\mathfrak{p}$ = the ideal of nonunits of $\mathfrak{o}$,
$\mathfrak{O}$ = the ring of integers of K,
$\mathfrak{P}$ = the ideal of nonunits of $\mathfrak{O}$,
$\bar{k} = \mathfrak{o}/\mathfrak{p}$, $\bar{K} = \mathfrak{O}/\mathfrak{P}$,
V_k = the value group of k,
V_K = the value group of K,
$e = e(K/k) =$ the index $(V_K:V_k)$,
$f = f(K/k) = [\bar{K}:\bar{k}]$.

The degree f is called the *residue class degree* of K/k and e is called the *ramification* of K/k.

If $a \in \mathfrak{O}$ we shall denote its image in $\bar{K}$ by $\bar{a}$. We shall also denote by $u_1, \ldots, u_s$ elements of $\mathfrak{O}$ such that $\bar{u}_1, \ldots, \bar{u}_s$ are linearly independent over $\bar{k}$, and by $c_1, \ldots, c_t$ elements of K^* such that $|c_1|, \ldots, |c_t|$ represent distinct cosets of V_k in V_K.

LEMMA. *For any $a_1, \ldots, a_s \in k$,*

$$|a_1 u_1 + \cdots + a_s u_s| = \max_{1 \leqslant i \leqslant s} |a_i|.$$

Proof. We may suppose that $|a_1| = \max\limits_{1 \leqslant i \leqslant s} |a_i|$. If $|a_1| = 0$, then $a_i = 0$ for all i, and therefore the lemma is true in this case. Suppose that $|a_1| \neq 0$.

Then $a_1 \neq 0$ and we have

$$|a_1 u_1 + \cdots + a_s u_s| = |a_1| \, |u_1 + a_1^{-1} a_2 u_2 + \cdots + a_1^{-1} a_s u_s|.$$

For $i = 2, \ldots, s$ we have $|a_1^{-1} a_i| \leqslant 1$ and so $|u_1 + a_1^{-1} a_2 u_2 + \cdots + a_1^{-1} a_s u_s| \leqslant 1$. If this inequality is strict then $\bar{u}_1 + \overline{a_1^{-1} a_2} \bar{u}_2 + \cdots + \overline{a_1^{-1} a_s} \bar{u}_s = 0$, which contradicts our choice of $u_1, \ldots, u_s$. Hence we must have equality and the lemma follows. $\|$

THEOREM 1. $ef \leqslant n$.

Proof. It is enough to show that $st \leqslant n$. We shall show that the st elements of K, $c_i u_j$, $i = 1, \ldots, t$, $j = 1, \ldots, s$, are linearly independent over k. Let $a_{ij} \in k$ for all i and j. We have for $i = 1, \ldots, t$.

$$\left| \sum_{j=1}^{s} a_{ij} c_i u_j \right| = |c_i| \left| \sum_{j=1}^{s} a_{ij} u_j \right|$$
$$= |c_i| \max_{i \leqslant j \leqslant s} |a_{ij}|.$$

Since the $|c_i|$ represent distinct cosets of V_k in V_K the non-zero ones among these values are distinct. Hence

$$\left| \sum_{i=1}^{t} \sum_{j=1}^{s} a_{ij} c_i u_j \right| = \max_{1 \leqslant i \leqslant t} |c_i| \max_{1 \leqslant j \leqslant s} |a_{ij}|$$
$$= \max_{\substack{1 \leqslant i \leqslant t \\ 1 \leqslant j \leqslant s}} |c_i| \, |a_{ij}|.$$

Now suppose that $\sum_{i=1}^{t} \sum_{j=1}^{s} a_{ij} c_i u_j = 0$. Then $|c_i| \, |a_{ij}| = 0$ for all i and j. Since $c_i \neq 0$ for all i this implies that $a_{ij} = 0$ for all i and all j. $\|$

We now consider the question of when ef is actually equal to n. We shall not give a complete answer to this question, but we will determine one important case in which we do have $ef = n$.

We define an extended real-valued function on k (or on K) as follows: let c be a real number greater than one and set

$$Va = -\log_c |a|.$$

Then

   (i) $V0 = \infty$ and $Va = \infty$ only if $a = 0$,

  (ii) $Vab = Va + Vb$,

 (iii) $V(a + b) \geqslant \min(Va, Vb)$.

The function V induces an isomorphism from V_k onto a subgroup of the additive group of real numbers. The subgroups of the additive group of real numbers fall into two categories, those which form a discrete subset of the real numbers and those which form a dense subset of the real numbers (with respect to the Euclidean topology of the real line). If the isomorphism determined by V maps V_k onto a discrete subgroup of the additive group of real numbers we say that $|\ |$ is a *discrete valuation* and we call k a *discrete valuated field*.

Suppose that k is a discrete valuated field. Then there is an element $\pi \in k$ such that $V\pi = \min \{Va \,|\, a \in k$ and $Va > 0\}$. The element π is also characterized by the fact that $|\pi|$ is the largest element of V_k which is less than one.

PROPOSITION. *If $a \in k^*$ then Va is an integral multiple of $V\pi$.*

Proof. Suppose that Va is not an integral multiple of $V\pi$. If Va is negative we replace a by a^{-1}. Hence we may assume that Va is positive. Let r be the positive integer such that $(r - 1)V\pi < Va < rV\pi$. Then $0 < Va - (r - 1)V\pi < V\pi$ or $0 < Va\pi^{1-r} < V\pi$, which is contrary to our choice of π. ‖

If $a \in k^*$ it follows from the proposition that $|a| = |\pi|^n$ for some integer n. Then $|a\pi^{-n}| = 1$ and so $a\pi^{-n} = u$ where u is a unit in k. Thus $a = u\pi^n$. This means that π is a prime element of the ring $\mathfrak{o}$. If π_1 is another prime element of $\mathfrak{o}$ then $\pi_1 = u\pi$ where u is a unit and therefore $\mathfrak{o}$ is a ring with essentially only one prime. We call π a *prime element* of the valuated field k. From now on we shall use $|\pi|^{-1}$ for the base c of the logarithm in the definition of the function V. If $a = u\pi^n$ with u a unit then $Va = n$. We note that Va is independent of the choice of the particular prime element π. With this specification of the value of c the function V is called the *ordinal function* of k and we shall write ord a instead of Va.

DEFINITION.   A complete discrete non-Archimedean valuated field is called a *local field*.

LEMMA.   *Let k be a local field with valuation $|\ |$. For each integer m let π_m be an element of k such that ord $\pi_m = m$. Let $\mathfrak{R}$ be a complete set of representatives in $\mathfrak{o}$ of the elements of $\bar{k}$, that is, $\mathfrak{R}$ consists of exactly one element from each of the residue classes of $\mathfrak{p}$ in $\mathfrak{o}$. Then every $a \in k^*$ can be written uniquely in the form*

$$a = \sum_{i=n}^{\infty} c_i \pi_i,$$

where $n = $ ord a, $c_i \in \mathfrak{R}$ for each i, and $c_n \notin \mathfrak{p}$.

Proof. If ord $a = n$ then $|a/\pi_n| = 1$ and so a/π_n is a unit in $\mathfrak{o}$. Hence there is an element $c_n \in \mathfrak{R}$, $c_n \notin \mathfrak{p}$, such that $|a/\pi_n - c_n| < 1$ or $|a - c_n\pi_n| < |\pi_n|$. Thus $a = c_n\pi_n + a_1$ where $|a_1| < |\pi_n|$. Then ord $a_1 > $ ord $\pi_n = n$, and since ord a_1 is an integer we must have ord $a_1 \geqslant n + 1 = $ ord π_{n+1}, that is, $|a_1/\pi_{n+1}| \leqslant 1$. So there is an element $c_{n+1} \in \mathfrak{R}$ such that

$$|a_1/\pi_{n+1} - c_{n+1}| < 1 \text{ or } |a_1 - c_{n+1}\pi_{n+1}| < |\pi_{n+1}|.$$

Thus $a_1 = c_{n+1}\pi_{n+1} + a_2$ where $|a_2| < |\pi_{n+1}|$, and $a = c_n\pi_n + c_{n+1}\pi_{n+1} + a_2$. If we repeat this argument a number of times we obtain

$$a = \sum_{i=n}^{n+h} c_i\pi_i + a_{h+1}, \qquad |a_{h+1}| < |\pi_{h+1}|.$$

Then $a_{h+1} \to 0$ as $h \to \infty$ and so $a = \sum_{i=n}^{\infty} c_i\pi_i$.

Now assume that we also have $a = \sum_{i=n}^{\infty} d_i\pi_i$ where $d_i \in \Re$ for each i, and $d_n \notin \mathfrak{p}$. Then $0 = \sum_{i=n}^{\infty} (c_i - d_i)\pi_i$. Suppose $c_i \neq d_i$ for some i and let m be the smallest integer such that $c_m \neq d_m$. Then $0 = \sum_{i=m}^{\infty} (c_i - d_i)\pi_i$ and ord $((c_m - d_m)\pi_m) \geqslant m + 1$, which implies that ord $(c_m - d_m) \geqslant 1$. Thus c_m and d_m belong to the same residue class of $\mathfrak{p}$ in $\mathfrak{o}$, and so must actually be equal, which contradicts our choice of m. Therefore, we must have $c_i = d_i$ for all i. $\parallel$

Let k be a local field and let K be a finite extension of k. Since V_k has finite index in V_K (it follows from Theorem 1 that both e and f are finite) the valuation on K is also discrete and so K is also a local field. Let π be a prime element of k and Π a prime element of K. The value group of k is an infinite cyclic group generated by $|\pi|$ while that of K is an infinite cyclic group generated by $|\Pi|$. Since $(V_K : V_k) = e$, $|\Pi|^e$ is a generator of V_k. Hence $|\Pi|^e = |\pi|$ and each element of V_K can be written uniquely in the form $|\pi^i\Pi^j|$ where $-\infty < i < \infty$ and $0 \leqslant j \leqslant e - 1$. Now let $u_1, \ldots, u_f$ be elements of $\mathfrak{O}$ such that $\bar{u}_1, \ldots, \bar{u}_f$ form a basis of $\bar{K}$ over $\bar{k}$. Then every element of $\bar{K}$ has a representative in $\mathfrak{O}$ of the form $c_1u_1 + \cdots + c_fu_f$ where $c_1, \ldots, c_f \in \mathfrak{o}$. Hence, by the lemma, if $a \in K^*$ there are elements $c_{hij} \in \mathfrak{o}$ such that

$$a = \sum_h \sum_{i=1}^{f} \sum_{j=0}^{e-1} c_{hij}u_i\pi^h\Pi^j$$

where h runs from some integer to ∞. Then

$$a = \sum_{i=1}^{f} \sum_{j=0}^{e-1} \left(\sum_h c_{hij}\pi^h\right) u_i\Pi^j.$$

This change of order of summation is allowed because k is complete and so the inner sum (on h) converges to some element of k for each i and j. Thus the ef elements of K, $u_i\Pi^j$, $i = 1, \ldots, f$, $j = 0, \ldots, e - 1$, span K as a vector space over k. Therefore $n \leqslant ef$. Combining this with Theorem 1 we have

THEOREM 2.   *If k is a local field and if K is a finite extension of k then $ef = n$.*

Throughout the remainder of this chapter and the next, all valuations on the base field, usually denoted by k, which are non-Archimedean will be assumed to be discrete. Prolongations of these valuations to valuations on finite extensions of k will then be discrete. We note that the p-adic valuations on the field Q are discrete with p as prime element. Also, the valuations on a field of rational functions $k(x)$ which are trivial on k are discrete. Since the value group of a completion of a discrete valuated field k is the same as the value group of k, a prime element of k is also a prime element of the completion of k. Hence the prime p is a prime element of the field of p-adic numbers.

2. Unramified and tamely ramified extensions

Let k be a local field with valuation $|\ |$. Let $\mathfrak{o}$ be the ring of integers of k, $\mathfrak{p}$ the ideal of nonunits of $\mathfrak{o}$, and π a prime element of k. If K is a finite extension of k it follows from the corollary to Theorem 11 in Chapter 3 that $|\ |$ has a unique prolongation to a valuation on K which we also denote by $|\ |$. We shall denote the ring of integers of the valuated field K by $\mathfrak{O}_K$, the ideal of nonunits of $\mathfrak{O}_K$ by $\mathfrak{P}_K$, and a prime element of K by Π_K, or we shall use these same letters without the subscript. If K is an arbitrary algebraic extension of k we use the same notations, but we observe that if K/k is not finite then the valuation on K may not be discrete. As previously we denote the residue class field of K by $\bar{K}$.

Let C be an algebraic closure of k: all of the extensions of k which we shall consider will be subfields of C.

PROPOSITION. *The field $\bar{C}$ is an algebraic closure of $\bar{k}$.*

Proof. We shall first show that $\bar{C}/\bar{k}$ is algebraic. Let a be an arbitrary element of $\mathfrak{O}_C$ so that $\bar{a}$ is an arbitrary element of $\bar{C}$. Let $f(x) = \mathrm{Irr}\,(k,a) = a_0 + a_1 x + \cdots + a_n x^n$ and let $b \in k$ be such that $|b| = \max\,(|a_0|, |a_1|, \ldots, |a_n|)$. Then $g(x) = b^{-1} f(x)$ has all of its coefficients in $\mathfrak{o}$ and at least one of its coefficients is a unit in $\mathfrak{o}$. Hence $\bar{g}(x) \neq 0$ and we have $\bar{g}(\bar{a}) = 0$ so that $\bar{a}$ is algebraic over $\bar{k}$. Therefore $\bar{C}/\bar{k}$ is algebraic.

To show that $\bar{C}$ is algebraically closed we shall show that every nonconstant monic irreducible polynomial $h(x) \in \bar{k}[x]$ splits in $\bar{C}[x]$ (the reader should satisfy himself that this is sufficient). Let $f(x)$ be a polynomial in $\mathfrak{o}[x]$ such that $\bar{f}(x) = h(x)$. We may choose $f(x)$ to be monic and to have the same degree as $h(x)$. Then $f(x)$ is irreducible in $k[x]$. For, if $f(x)$ could be factored in a nontrivial way in $k[x]$ then it could be factored in a nontrivial way in $\mathfrak{o}[x]$, and by passing to the residue class field we would obtain a nontrivial factorization of $h(x)$ in $\bar{k}[x]$. However, $f(x)$ splits in $C[x]$:

$$f(x) = (x - a_1) \cdots (x - a_n), \qquad a_1, \ldots, a_n \in C.$$

As we have seen in the last chapter, since $f(x)$ is irreducible in $k[x]$, all of the roots of $f(x)$ in C have the same valuation. Then since the constant term of $f(x)$ lies in $\mathfrak{o}$ we have $|a_i| \leqslant 1$ for $i = 1, \ldots, n$, that is, each $a_i \in \mathfrak{D}_C$. But then $h(x) = (x - \bar{a}_1) \cdots (x - \bar{a}_n)$, and so $h(x)$ splits in $\bar{C}[x]$. ∥

All of the extensions of $\bar{k}$ which we shall consider will be subfields of $\bar{C}$.

DEFINITION.   A finite extension K of k with ramification e is said to be *unramified* if $e = 1$ and $\bar{K}/\bar{k}$ is separable.

THEOREM 3.   *There is a one-one correspondence between the unramified extensions of k (in C) and the finite separable extensions of $\bar{k}$ (in $\bar{C}$).*

In proving this theorem we shall use an important result which is due to Krasner.

THEOREM 4.   *Let $a \in C$ be separable over k and let*

$$r = \min |\sigma(a) - a|,$$

where σ ranges over all k-isomorphisms of $k(a)$ into C other than the identity isomorphism. If $b \in C$ and if $|b - a| < r$ then $k(a) \subseteq k(b)$.

Proof.   Consider the extension $k(a,b)$ of $k(b)$. Since a is separable over k it is separable over $k(b)$. Therefore, by Theorem 20 in Chapter 1, there are exactly $[k(a,b): k(b)]$ $k(b)$-isomorphisms of $k(a,b)$ onto subfields of C. Let σ be one of these isomorphisms. The mapping σ induces a k-isomorphism of $k(a)$ into C which we also denote by σ. We have $\sigma(b - a) = b - \sigma(a)$, and conjugate elements in C have the same valuation since they have the same norm. Hence $|b - \sigma(a)| = |b - a| < r$. Then

$$|\sigma(a) - a| = |(b - \sigma(a)) - (b - a)| \leqslant \max(|b - \sigma(a)|, |b - a|) < r$$

which implies that $\sigma(a) = a$. Thus σ is the identity isomorphism of $k(a,b)$. It follows that $[k(a,b): k(b)] = 1$ and so $a \in k(b)$. ∥

Proof of Theorem 3.   Consider a finite separable extension of $\bar{k}$. By Theorem 24 of Chapter 1 this extension has a primitive element and so is equal to $\bar{k}(\bar{b})$ for some $\bar{b} \in \bar{C}$. Let $f(x)$ be a monic polynomial in $\mathfrak{o}[x]$ such that $\bar{f}(x) = \text{Irr}\,(\bar{k},\bar{b})$. If $f(x)$ factors in a nontrivial way in $k[x]$ then it factors in a nontrivial way in $\mathfrak{o}[x]$ and it follows that $\bar{f}(x)$ factors in a nontrivial way in $\bar{k}[x]$. But $\bar{f}(x)$ is irreducible in $\bar{k}[x]$ so that $f(x)$ must be irreducible in $k[x]$. Let b be a root of $f(x)$ in C and let $K = k(b)$. It is clear that we can choose b to be a representative of $\bar{b}$. Let $n = [K: k] = [\bar{k}(\bar{b}): \bar{k}] = \deg \bar{f}(x) = \deg f(x)$. We have $\bar{b} \in \bar{K}$ and so $\bar{k}(\bar{b}) \subseteq \bar{K}$. Hence $n \leqslant [\bar{K}: \bar{k}] \leqslant [K: k] = n$ which implies that $[\bar{K}: \bar{k}] = [K: k]$. Therefore, the ramification of K over k is one and $\bar{K} = \bar{k}(\bar{b})$, which is separable over $\bar{k}$.

There is a mapping from the family of unramified extensions of k into the family of separable extensions of $\bar{k}$, namely, the mapping taking K onto $\bar{K}$.

We have just shown that this mapping is *onto*, and we shall now show that it is *one-one*. In the notation of the preceding paragraph we shall show that K is the only unramified extension of k in C having $\bar{k}(\bar{b})$ as its residue class field. Let F be an unramified extension of k in C such that $\bar{F} = \bar{K} = \bar{k}(\bar{b})$. Let $b_1, \ldots, b_n$ be the roots of $f(x)$ in C. Then the roots of $\bar{f}(x)$ in $\bar{C}$ are $\bar{b}_1, \ldots, \bar{b}_n$, and these are distinct elements of $\bar{C}$ since $\bar{b}$ is separable over $\bar{k}$. Hence $|b - b_i| = 1$ for $i = 2, \ldots, n$. Let $a \in F$ be such that $\bar{a} = \bar{b}$. Then $|a - b| < 1$ and, by Theorem 4, we have $k(b) \subseteq k(a) \subseteq F$, that is, $K \subseteq F$. But K and F both have ramification one over k and they have the same residue class degree. Hence $[K : k] = [F : k]$ and therefore $F = K$. ‖

COROLLARY. *Let K and F be unramified extensions of k in C. If $\bar{K} \subseteq \bar{F}$ then $K \subseteq F$.*

Proof. Since $\bar{F}/\bar{K}$ is separable there is an unramified extension L of K in C such that $\bar{L} = \bar{F}$. Then L/k is unramified and so $L = F$. Therefore $K \subseteq F$. ‖

Let K be a finite extension of k and let T_0 be the separable closure of $\bar{k}$ in $\bar{K}$ (see Corollary 4 to Theorem 10 in Chapter 1). By the argument used in the proof of Theorem 3 we can show that there is an unramified extension T of k such that $k \subseteq T \subseteq K$ and $\bar{T} = T_0$. If F is an unramified extension of k in K then $\bar{F} \subseteq \bar{T}$ and so by the above corollary, $F \subseteq T$. Therefore, T is maximum among the unramified extensions of k which are contained in K. The field T is called the *inertia field* of the extension K of k.

DEFINITION. A finite extension K of k with ramification e is said to be *tamely ramified* if $\bar{K}/\bar{k}$ is separable and if e is not divisible by char $\bar{k}$.

PROPOSITION 1. *Let K be a tamely ramified extension of k with ramification e. Let T be the inertia field of K. Then there is a prime element π of T such that $K = T(\sqrt[e]{\pi})$.*

Proof. Let Π be a prime element of K. Then $|\Pi|^e$ is a generator of the value group of T and so there is a prime element π_1 of T and a unit u of K such that $\Pi^e = u\pi_1$. Since $\bar{K} = \bar{T}$ there is a unit v of T such that $\bar{v} = \bar{u}$: then $|u - v| < 1$. Then $\pi = v\pi_1$ is a prime element of T and if we set $c = \pi_1(u - v)$ we have $|c| < |\pi|$ and $\Pi^e = \pi + c$. Now set $f(x) = x^e - \pi$. By the Eisenstein irreducibility criterion, which is proved below, $f(x)$ is irreducible in $T[x]$. The result will follow if we can show that K contains a root of $f(x)$.

Since e is not divisible by char $\bar{k}$, it is not divisible by char k. Hence $f(x)$ has distinct roots in an algebraic closure C of k which contains K: let

$$f(x) = (x - a_1) \cdots (x - a_e), \qquad a_1, \ldots, a_e \in C.$$

By Theorem 11 in Chapter 3, $|a_i| = \sqrt[e]{|\pi|} = |\Pi|$ for each i, and so $|a_i - a_j| \leqslant |\Pi|$ for all i and j. If we consider e as an element of k then $|e| = 1$. Hence

$$|\Pi|^{e-1} = |a_i^{e-1}| = |ea_i^{e-1}| = |f'(a_i)| =$$
$$|(a_i - a_1) \cdots (a_i - a_{i-1})(a_i - a_{i+1}) \cdots (a_i - a_e)|,$$

and so we must have $|a_i - a_j| = |\Pi|$ for all i and j with $i \neq j$. Also, $|(\Pi - a_1) \cdots (\Pi - a_e)| = |f(\Pi)| = |\Pi^e - \pi| = |c| < |\pi| = |\Pi|^e$ which means that for some i we have $|\Pi - a_i| < |\Pi|$. Hence, by Theorem 4, we have $k(a_i) \subseteq k(\Pi) \subseteq K$, and the proposition is proved. ‖

PROPOSITION 2. *Let n be a positive integer not divisible by char $\bar{k}$. Let $b \in k$ and let $K = k(a)$ where a is a root of $x^n - b$. Then K is a tamely ramified extension of k.*

Proof. Since $|a|^n = |b|$ is in the value group of k there is a smallest positive integer d such that $|a|^d$ is in the value group of k. Let $|a|^d = |b_1|$ where $b_1 \in k$ and set $c = a^d b_1^{-1}$. Then $|c| = 1$ and $c^{n/d} = bb_1^{-n/d} = c_1 \in k$. Hence c is a root of $g(x) = x^{n/d} - c_1$ where $|c_1| = 1$. Since $\bar{g}'(x) \neq 0$ it has distinct roots in an algebraic closure of $\bar{k}$, and so by Exercise 12, $k(c)$ is an unramified extension of k. Note that a is a root of $x^d - b_1 c \in k(c)[x]$. Since the value groups of $k(c)$ and k are the same, d is the smallest positive integer such that $|a|^d$ is in the value group of $k(c)$. Hence $d \leqslant e(K/k(c))$. However, $[K: k(c)] \leqslant d$ and so $e(K/k(c)) = [K: k(c)] = d$. Thus $f(K/k(c)) = 1$ and $\bar{K} = \overline{k(c)}$ is separable over $\bar{k}$. Therefore, K is a tamely ramified extension of k. ‖

Let K be a finite extension of k and let T be the inertia field of K. Then T is a tamely ramified extension of k. Since a tamely ramified extension of a tamely ramified extension of k is itself a tamely ramified extension of k, it is clear that there is a maximal tamely ramified extension V of k in K such that $T \subseteq V$. Let F be any other tamely ramified extension of k in K. Then FV/V is tamely ramified (see Exercise 14) and so FV/k is tamely ramified. Hence $FV = V$ and so $F \subseteq V$. Therefore, V contains all subfields of K which are tamely ramified extensions of k.

If char $\bar{k} = p$ we let $e = dp^r$ where p does not divide d: if char $\bar{k} = 0$ we replace p^r by 1. If e_0 is the ramification of V/k then e_0 divides e but is not divisible by char $\bar{k}$. Hence e_0 divides d and we write $d = e_0 t$. If Π is a prime element of K then $|\Pi^{p^r}|^t$ is a generator of the value group of V and so $(\Pi^{p^r})^t = u\pi_1$ where π_1 is a prime element of V and u is a unit in K. We can now argue as in the proof of Proposition 1 to show that $V(a) \subseteq K$ where a is a root of $x^t - \pi$, where π is a prime element of V, and where this polynomial is irreducible in $V[x]$. If $\bar{K} = \bar{V}$, the argument is exactly the same as in the proof of Proposition 1: if $\bar{K} \neq \bar{V}$ then some modification in the argument must be made which takes advantage of the fact that $\bar{K}/\bar{V}$ is purely inseparable.

By Proposition 2, $V(a)/V$ is tamely ramified. Hence we must have $t = 1$ and $e_0 = d$. If we now apply Proposition 1 we see that we have proved

THEOREM 5.   *There is a prime element π of T such that $V = T(\sqrt[d]{\pi})$ is a tamely ramified extension of k in K which contains all subfields of K which are tamely ramified extensions of k.*

The field V is called the *ramification field* of the extension K of k.

THEOREM 6.   (Eisenstein Irreducibility Criterion.)   *Let k be a non-Archimedean valued field (not necessarily complete). Let*

$$f(x) = x^n + a_{n-1}x^{n-1} + \cdots + a_1 x + a_0 \in k[x].$$

Assume that ord $a_i \geqslant 1$ *for* $i = 1, \ldots, n - 1$ *and* ord $a_0 = 1$. *Then $f(x)$ is irreducible in $k[x]$.*

Proof.   If we can factor $f(x)$ in $k[x]$ then we can factor it in $\mathfrak{o}[x]$, where $\mathfrak{o}$ is the ring of integers of k. Suppose that

$$f(x) = (x^r + b_{r-1}x^{r-1} + \cdots + b_0) \cdot (x^s + c_{s-1}x^{s-1} + \cdots + c_0),$$

where all the b's and all the c's are in $\mathfrak{o}$, and let $b_r = c_s = 1$. Since ord $a_0 =$ ord $b_0 c_0 = $ ord $b_0 +$ ord $c_0 = 1$ we have ord $b_0 = 0$ or ord $c_0 = 0$. Assume that ord $b_0 = 0$: then ord $c_0 = 1$. Let ord $c_i \geqslant 1$ for $i = 0, \ldots, h$ and ord $c_{h+1} = 0$ (we certainly have $h \leqslant s - 1$). Now consider the coefficient a_{h+1} of x^{h+1} in $f(x)$:

$$a_{h+1} = b_0 c_{h+1} + b_1 c_h + \cdots + b_{h+1}c_0.$$

Every term on the right-hand side has ordinal $\geqslant 1$ except $b_0 c_{h+1}$, which has ordinal 0. Thus ord $a_{h+1} = 0$ which implies that $h = n - 1$. Thus $n \leqslant s \leqslant n$, that is, $s = n$ and so $r = 0$ and the first of the two factors of $f(x)$ is a constant. ‖

To summarize the results of this section, let k be a local field and let K be a finite extension of k. Then $k \subseteq T \subseteq V \subseteq K$ where T/k is unramified and V/k (and so V/T) is tamely ramified. Also $\bar{T}$ is the separable closure of $\bar{k}$ in $\bar{K}$ and so $[T:k] = [\bar{T}:\bar{k}] = [\bar{K}:\bar{k}]_s$. We have $\bar{V} = \bar{T}$ and $[K:V] = e(K/V)[\bar{K}:\bar{V}]$, both of which are powers of p if char $\bar{k} = p$ and equal to one if char $\bar{k} = 0$. In many cases of great interest the field $\bar{k}$ is perfect so that $\bar{T} = \bar{V} = \bar{K}$. Then K is *fully ramified* over V, that is, its degree over V is equal to its ramification over V.

We close this section with a result which is of great importance in algebraic number theory.

THEOREM 7.   *Suppose that $\bar{k}$ is a finite field. Then for each positive integer n there is one and only one unramified extension of degree n over k in C and it is obtained from k by adjoining a root of unity to k.*

Proof. First suppose that F and L are unramified extensions of k such that $[F:k] = [L:k] = n$. Then F and L have the same degree over $\bar{k}$ and so $\bar{F} = \bar{L}$. Then $F = L$ by Theorem 3. Since $\bar{k}$ has exactly one extension of degree n, and since $\bar{k}$ is perfect, k has exactly one unramified extension of degree n (in C).

Suppose that $\bar{k}$ has q elements and let $\bar{K}$ be the extension of $\bar{k}$ in $\bar{C}$ with $[\bar{K}:\bar{k}] = n$: then $\bar{K}$ has q^n elements. Let K be the corresponding unramified extension of k in C. The roots of $x^{q^n-1} - 1$ in $\bar{C}$ are the $q^n - 1$ elements of $\bar{K}$ different from zero. If ζ is a primitive $(q^n - 1)$th root of unity in C then $k(\zeta)$ is unramified over k (Exercise 12). Since $x^{q^n-1} - 1$ splits in $k(\zeta)[x]$ it splits in $\overline{k(\zeta)}[x]$ and $\bar{K} \subseteq \overline{k(\zeta)}$. By the corollary to Theorem 3 this implies that $K \subseteq k(\zeta)$. On the other hand, $\overline{k(\zeta)} \subseteq \bar{K}$ and we let $a \in K$ be such that $\bar{a} = \bar{\zeta}$. Then $|a - \zeta| < 1$. However, if ζ_1 and ζ_2 are distinct roots of $x^{q^n-1} - 1$ in C then $\bar{\zeta}_1$ and $\bar{\zeta}_2$ are distinct roots of $x^{q^n-1} - 1$ in $\bar{C}$ and so $|\zeta_1 - \zeta_2| = 1$. Therefore, by Theorem 4, $k(\zeta) \subseteq k(a) \subseteq K$. Thus $K = k(\zeta)$. ∥

3. The different

Let k be a local field and let $\mathfrak{o}$ be the ring of integers of k, $\mathfrak{p}$ the ideal of nonunits of $\mathfrak{o}$, and π a prime element of k. Then $\mathfrak{p}$ is a principal ideal of $\mathfrak{o}$ with π as a generator, that is, $\mathfrak{p} = \pi\mathfrak{o}$. For each positive integer n we have $\mathfrak{p}^n = \pi^n\mathfrak{o}$, and we set $\mathfrak{p}^0 = \mathfrak{o}$. The ideals $\mathfrak{p}^n(n \geqslant 0)$ are all of the nonzero ideals of $\mathfrak{o}$. To show this let $\mathfrak{a}$ be an ideal of $\mathfrak{o}$ other than the zero ideal and $\mathfrak{o}$ itself. Then $\mathfrak{a} \subseteq \mathfrak{p}$ and since $\bigcap\limits_{n=1}^{\infty} \mathfrak{p}^n = 0$ there is an integer n such that $\mathfrak{a} \subseteq \mathfrak{p}^n$ but $\mathfrak{a} \not\subseteq \mathfrak{p}^{n+1}$. Let $a \in \mathfrak{a}$ and $a \notin \mathfrak{p}^{n+1}$: then $a = \pi^n u$ where u is a unit. Let $b \in \mathfrak{p}^n$ and write $b = \pi^n c$ where $c \in \mathfrak{o}$. Then $b = a(u^{-1}c) \in \mathfrak{a}$ and so $\mathfrak{p}^n \subseteq \mathfrak{a}$. Hence $\mathfrak{a} = \mathfrak{p}^n$.

DEFINITION.   A nonempty subset A of k is called a *fractional ideal* of k if the following conditions hold:
  (i) if $a, b \in A$ then $a - b \in A$,
  (ii) if $a \in A$ and $c \in \mathfrak{o}$ then $ac \in A$,
  (iii) there is an element $d \in \mathfrak{o}$, $d \neq 0$, such that $dA \subseteq \mathfrak{o}$.

Let A be a nonzero fractional ideal of k and let $d \in \mathfrak{o}$, $d \neq 0$, be such that $dA \subseteq \mathfrak{o}$. Then dA is an ideal of $\mathfrak{o}$ and so $dA = \pi^n\mathfrak{o}$ for some integer n. If $\mathrm{ord}\, d = t$ then $A = \pi^{n-t}\mathfrak{o}$. Note that each ideal of $\mathfrak{o}$ is a fractional ideal of k. Therefore the nonzero fractional ideals of k are given by $\pi^n\mathfrak{o}$ for $-\infty < n < \infty$.

If A and B are fractional ideals of k we define their product to be

$$AB = \{\textstyle\sum a_i b_i \mid a_i \in A, b_i \in B\}$$

where only finite sums are considered. It is easily seen that AB is a fractional ideal of k. If $A = \pi^m \mathfrak{o}$ and $B = \pi^n \mathfrak{o}$ then $AB = \pi^{m+n} \mathfrak{o}$. Note that the set of fractional ideals of k different from the zero ideal forms an infinite cyclic group with respect to multiplication, and that this set is linearly ordered with $\pi^n \mathfrak{o} \subseteq \pi^m \mathfrak{o}$ if $n \geqslant m$. Furthermore, $\bigcup_{n=1}^{\infty} \pi^{-n} \mathfrak{o} = k$.

Let K be a finite separable extension of k: we employ the notation of Section 1. Set

$$\mathfrak{D}' = \{a \in K \mid T_{K/k}(ab) \in \mathfrak{o} \text{ for all } b \in \mathfrak{D}\}.$$

THEOREM 8. *The set $\mathfrak{D}'$ is a fractional ideal of K and $\mathfrak{D} \subseteq \mathfrak{D}'$.*

Proof. If $b, c \in \mathfrak{D}$ then $bc \in \mathfrak{D}$ and so $T_{K/k}(bc) \in \mathfrak{o}$. Thus $\mathfrak{D} \subseteq \mathfrak{D}'$. Since K/k is separable there is an $a \in K$ such that $T_{K/k}(a) = a' \neq 0$ (see Theorems 28 and 29 in Chapter 1). We may assume that $a \in \mathfrak{D}$ (why?). For any positive integer n we have $T_{K/k}(\pi^{-n}a) = \pi^{-n}a'$ and if we choose $n > \operatorname{ord} a'$ we will have $\pi^{-n}a' \notin \mathfrak{o}$. Thus $\pi^{-n} \notin \mathfrak{D}'$ and so $\mathfrak{D}' \neq K$.

Since $\mathfrak{D} \subseteq \mathfrak{D}' \subset K$ there is a non-negative integer m such that $\Pi^{-m}\mathfrak{D} \subseteq \mathfrak{D}'$ but $\Pi^{-m-1}\mathfrak{D} \nsubseteq \mathfrak{D}'$, where Π is a prime element of K. Let $a \in \mathfrak{D}'$ and write $a = \Pi^s v$ where v is a unit in $\mathfrak{D}$. Let $b = \Pi^{-m-1}u$ be in $\Pi^{-m-1}\mathfrak{D}$ but not in $\mathfrak{D}'$. If u is not a unit in $\mathfrak{D}$ then $b \in \Pi^{-m}\mathfrak{D} \subseteq \mathfrak{D}'$: hence u is a unit. Since $b \notin \mathfrak{D}'$ there is an element $c \in \mathfrak{D}$ such that $T_{K/k}(bc) \notin \mathfrak{o}$. Write $c = \Pi^t w$ where w is a unit in $\mathfrak{D}$ and $t \geqslant 0$. Then

$$T_{K/k}(\Pi^{t-m-1}uw) \notin \mathfrak{o}$$

or

$$T_{K/k}(\Pi^{t-m-1-s}v^{-1}uwa) \notin \mathfrak{o}.$$

But $a \in \mathfrak{D}'$ and so this implies that $\Pi^{t-m-1-s}v^{-1}uw \notin \mathfrak{D}$, and therefore $t - m - 1 - s < 0$ or $s \geqslant t - m \geqslant -m$. Thus $a = \Pi^{-m}\Pi^{s+m}v$ and $s + m \geqslant 0$, so $\Pi^{s+m}v \in \mathfrak{D}$. Hence $a \in \Pi^{-m}\mathfrak{D}$. We have shown that $\mathfrak{D}' \subseteq \Pi^{-m}\mathfrak{D}$ and so $\mathfrak{D}' = \Pi^{-m}\mathfrak{D}$. Therefore $\mathfrak{D}'$ is a fractional ideal. ∥

DEFINITION. Let $\mathfrak{D}' = \Pi^{-m}\mathfrak{D}$ where $m \geqslant 0$. The ideal

$$\mathfrak{D}_{K/k} = \Pi^m \mathfrak{D}$$

of $\mathfrak{D}$ is called the *different* of K/k.

In this section and the next we shall obtain some properties of the different, including the relation of this concept to those of unramified extension and tamely ramified extension.

THEOREM 9. *Let K be a finite extension of k and L a finite extension of K with L/k separable. Then*

$$\mathfrak{D}_{L/k} = \mathfrak{D}_{L/K}\mathfrak{D}_{K/k}.$$

Proof. If we define $\mathfrak{D}'_{L/k}$, $\mathfrak{D}'_{L/K}$, and $\mathfrak{D}'_{K/k}$ in the obvious way, it is sufficient to show that $\mathfrak{D}'_{L/k} = \mathfrak{D}'_{L/K}\mathfrak{D}'_{K/k}$. If $a \in \mathfrak{D}'_{L/k}$ then $T_{L/k}(ab) = T_{K/k}(T_{L/K}(ab)) \in \mathfrak{o}$ for all $b \in \mathfrak{D}_L$. Hence for all $c \in \mathfrak{D}_K$ we have

$$T_{K/k}(cT_{L/K}(ab)) = T_{K/k}(T_{L/K}(abc)) \in \mathfrak{o}$$

which implies that $T_{L/K}(ab) \in \mathfrak{D}'_{K/k} = \Pi_K^{-n}\mathfrak{D}_K$ for all $b \in \mathfrak{D}_L$, where n is a non-negative integer. Then $T_{L/K}(\Pi_K^n ab) \in \mathfrak{D}_K$ for all $b \in \mathfrak{D}_L$ and so $\Pi_K^n a \in \mathfrak{D}'_{L/K}$, or $a \in \Pi_K^{-n}\mathfrak{D}'_{L/K} \subseteq \Pi_K^{-n}\mathfrak{D}_K\mathfrak{D}'_{L/K} = \mathfrak{D}'_{K/k}\mathfrak{D}'_{L/K}$. Therefore we have $\mathfrak{D}'_{L/k} \subseteq \mathfrak{D}'_{L/K}\mathfrak{D}'_{K/k}$.

On the other hand, suppose that $a \in \mathfrak{D}'_{L/K}\mathfrak{D}'_{K/k}$. We wish to show that $a \in \mathfrak{D}'_{L/k}$ and in doing this we may assume that $a = bc$ where $b \in \mathfrak{D}'_{K/k}$ and $c \in \mathfrak{D}'_{L/K}$. Let $d \in \mathfrak{D}_L$: then $T_{L/K}(cd) \in \mathfrak{D}_K$. Hence

$$T_{L/k}(ad) = T_{K/k}(T_{L/K}(ad)) = T_{K/k}(bT_{L/K}(cd)) \in \mathfrak{o}.$$

Hence $a \in \mathfrak{D}'_{L/k}$. Therefore $\mathfrak{D}'_{L/K}\mathfrak{D}'_{K/k} \subseteq \mathfrak{D}'_{L/k}$. ‖

THEOREM 10. *Let K be a finite separable extension of the local field k. Then K is an unramified extension of k if and only if $\mathfrak{D}_{K/k} = \mathfrak{D}$.*

Proof. First we shall show that if K/k is unramified then $\mathfrak{D}' = \mathfrak{D}$. We already know that $\mathfrak{D} \subseteq \mathfrak{D}'$ and so we must show that $\mathfrak{D}' \subseteq \mathfrak{D}$. We shall show that if $a \in K$ and $a \notin \mathfrak{D}$ then $a \notin \mathfrak{D}'$. Since K/k is unramified, a prime element of k is also a prime element of K and $\bar{K}/\bar{k}$ is separable. Hence there is an element $b \in \mathfrak{D}$ such that $T_{\bar{K}/\bar{k}}(\bar{b}) \neq 0$ so that $T_{K/k}(b) \notin \mathfrak{P}$. Since $a \notin \mathfrak{D}$ we have $a = \pi^{-n}u$ where π is a prime element of k, n is a positive integer, and u is a unit of $\mathfrak{D}$. Then $T_{K/k}(abu^{-1}) = T_{K/k}(\pi^{-n}b) = \pi^{-n}T_{K/k}(b) \notin \mathfrak{o}$. Hence $a \notin \mathfrak{D}'$.

Now suppose that K is not unramified over k. Let e be the ramification of K/k and let π be a prime element of k. There are two things which can cause K to fail to be unramified over k. It may be that $e > 1$ or it may be that $\bar{K}/\bar{k}$ is inseparable.

Suppose that $e > 1$. If $a \in \mathfrak{P}$ then $T_{K/k}(a) \in \mathfrak{p}$. If $b \in \mathfrak{D}$ then $ab \in \mathfrak{P}$ and $T_{K/k}(\pi^{-1}ab) = \pi^{-1}T_{K/k}(ab) \in \pi^{-1}\mathfrak{p} = \mathfrak{o}$. Hence $\pi^{-1}a \in \mathfrak{D}'$ for all $a \in \mathfrak{P}$. In particular, if Π is a prime element of K then $\pi^{-1}\Pi \in \mathfrak{D}'$. But ord $\pi^{-1}\Pi = -e + 1 < 0$ which implies that $\pi^{-1}\Pi \notin \mathfrak{D}$. Therefore $\mathfrak{D} \neq \mathfrak{D}'$.

Now suppose that $e = 1$ but that $\bar{K}/\bar{k}$ is inseparable. Then π is a prime element of K and we shall show that $\pi^{-1} \in \mathfrak{D}'$. Since $\pi^{-1} \notin \mathfrak{D}$ this will complete the proof. Let $a \in \mathfrak{D}$. By Theorems 28 and 29 in Chapter 1 we have $T_{K/k}(\bar{a}) = 0$ so that $T_{K/k}(a) \in \mathfrak{p}$. Hence $T_{K/k}(\pi^{-1}a) = \pi^{-1}T_{K/k}(a) \in \mathfrak{o}$. Therefore $\pi^{-1} \in \mathfrak{D}'$. ‖

COROLLARY. *Let K be a finite separable extension of a local field k and let T be the inertia field of K. Then $\mathfrak{D}_{K/k} = \mathfrak{D}_{K/T}$.*

Let K be a finite separable extension of the local field k with $[K:k] = n$ and let $a_1, \ldots, a_n$ be a basis of K/k. Consider the system of n equations in n unknowns

$$\sum_{j=1}^{n} T_{K/k}(a_i a_j) x_j = b_i, \qquad i = 1, \ldots, n,$$

where $b_1, \ldots, b_n \in k$. The determinant of this system of equations is the discriminant of the basis $a_1, \ldots, a_n$ and so is not zero by Theorem 29 in Chapter 1. Thus the system of equations has a unique solution $c_1, \ldots, c_n$ in k. If $c = c_1 a_1 + \cdots + c_n a_n$ then c is the unique solution of the n equations $T_{K/k}(a_i x) = b_i$, $i = 1, \ldots, n$.

For $j = 1, \ldots, n$ let $a_j{}'$ be the unique solution in K of the n equations $T_{K/k}(a_i x) = \delta_{ij}$. We shall show that $a_1{}', \ldots, a_n{}'$ are linearly independent over k. Suppose that $d_1 a_1{}' + \cdots + d_n a_n{}' = 0$ where $d_1, \ldots, d_n \in k$. Then for $i = 1, \ldots, n$,

$$0 = T_{K/k}(a_i(d_1 a_1{}' + \cdots + d_n a_n{}'))$$

$$= \sum_{j=1}^{n} d_j T_{K/k}(a_i a_j{}') = d_i.$$

Thus $a_1{}', \ldots, a_n{}'$ form a basis of K/k which is called the *complementary basis* of the basis $a_1, \ldots, a_n$.

If K is a finite extension of k (not necessarily separable over k) then a basis $a_1, \ldots, a_n$ of K/k is called a *minimal basis* if $a_1, \ldots, a_n \in \mathfrak{D}$ and if

$$\mathfrak{D} = a_1 \mathfrak{o} + \cdots + a_n \mathfrak{o}.$$

Here $a_i \mathfrak{o} = \{a_i b \mid b \in \mathfrak{o}\}$. Let e be the ramification of K/k, let Π be a prime element of K, and let $u_1, \ldots, u_f$ be elements of $\mathfrak{D}$ such that $\bar{u}_1, \ldots, \bar{u}_f$ form a basis of $\bar{K}/\bar{k}$. As we have seen in the proof of Theorem 2, the ef elements $u_i \Pi^j$, $i = 1, \ldots, f, j = 0, \ldots, e - 1$, form a basis of K/k. In fact, they form a minimal basis (Exercise 5). The importance of minimal bases is seen in the following result.

THEOREM 11. *Let K be a finite separable extension of the local field k and $a_1, \ldots, a_n$ be a minimal basis of K/k. If $a_1{}', \ldots, a_n{}'$ is the complementary basis of this basis then*

$$\mathfrak{D}' = a_1{}' \mathfrak{o} + \cdots + a_n{}' \mathfrak{o}.$$

Proof. Let $a \in K$ and write $a = b_1 a_1{}' + \cdots + b_n a_n{}'$ where $b_1, \ldots, b_n \in k$. Suppose that $a \in \mathfrak{D}'$. Then for $i = 1, \ldots, n$, we have

$$T_{K/k}(b_1 a_i a_1{}' + \cdots + b_n a_i a_n{}') = b_i \in \mathfrak{o}.$$

On the other hand, suppose that $b_1, \ldots, b_n \in \mathfrak{o}$. Then, if $c \in \mathfrak{D}$ and we write

$c = c_1 a_1 + \cdots + c_n a_n$ where $c_1, \ldots, c_n \in \mathfrak{o}$, we have

$$T_{K/k}(ac) = T_{K/k}\left(\sum_{i=1}^{n} \sum_{j=1}^{n} c_i b_j a_i a_j' \right)$$

$$= \sum_{i=1}^{n} c_i b_i \in \mathfrak{o}.$$

Thus $a \in \mathfrak{O}'$. ∥

4. Extensions K/k with $\bar{K}/\bar{k}$ separable

Let k be a local field and let K be a finite separable extension of k. In many instances of great interest $\bar{K}/\bar{k}$ is also separable. Often this is the only case that can occur: for example, when $\bar{k}$ is a finite field. In this section we shall study some consequences of this fact.

THEOREM 12. *If $\bar{K}/\bar{k}$ is separable then there is a minimal basis of K/k of the form $1, a, \ldots, a^{n-1}$ where $n = [K:k]$.*

Proof. Since $\bar{K}/\bar{k}$ is separable $\bar{K}$ has a primitive element with respect to $\bar{k}$. Hence there is an element $b \in \mathfrak{O}$ such that $1, \bar{b}, \ldots, \bar{b}^{f-1}$ is a basis of $\bar{K}/\bar{k}$. Then by remarks made in the last section the elements $b^i \Pi^j$, $i = 0, \ldots, f-1, j = 0, \ldots, e-1$, form a minimal basis of K/k, where Π is a prime element of K. Let T be the inertia field of K. If $T = k$, that is, if $f = 1$, then $1, \Pi, \ldots, \Pi^{e-1}$ is the desired minimal basis of K/k. If $T = K$, that is, if $e = 1$, then $1, b, \ldots, b^{f-1}$ is the desired minimal basis of K/k. It remains to consider the case when both $e > 1$ and $f > 1$.

Let $f(x) \in \mathfrak{o}[x]$ be such that $\bar{f}(x) = \text{Irr}(\bar{k}, \bar{b})$. Since $\bar{b}$ is separable over $\bar{k}, \bar{f}'(\bar{b}) \neq 0$ which means that $f'(b) \notin \mathfrak{P}$ (we continue to use the notation of Section 1). Now $\bar{f}(\bar{b}) = 0$ so that $f(b) \in \mathfrak{P}$ and we write $f(b) = \Pi^r u$ where u is a unit of $\mathfrak{O}$. Suppose $r > 1$. Then $f(b + \Pi) = f(b) + \Pi f'(b) + c\Pi^2$ for some $c \in \mathfrak{O}$. Since $f(b) \in \mathfrak{P}^2$ and $c\Pi^2 \in \mathfrak{P}^2$ but $\Pi f'(b) \notin \mathfrak{P}^2$ this implies that $f(b + \Pi) \notin \mathfrak{P}^2$. But $\overline{b + \Pi} = \bar{b}$ and so we may replace b by $b + \Pi$, call it b, and assume that $f(b) \in \mathfrak{P}$ but $f(b) \notin \mathfrak{P}^2$. Then $f(b)$ is a prime element of K and so the elements $b^i f(b)^j$, $i = 0, \ldots, f-1, j = 0, \ldots, e-1$, form a minimal basis of K/k. Thus every element of $\mathfrak{O}$ can be written in the form

$$\sum_{i=0}^{f-1} \sum_{j=0}^{e-1} c_{ij} b^i f(b)^j, \qquad c_{ij} \in \mathfrak{o}.$$

If $g(x) = \text{Irr}(k,b)$ then $\deg g(x) \leqslant n$ and so if $m \geqslant n$ we can multiply $g(b) = 0$ by a suitable power of b and obtain b^m as a linear combination of smaller powers of b: in fact, by repeating this we can express b^m as a linear combination of $1, b, \ldots, b^{n-1}$. The coefficients of this linear combination are in $\mathfrak{o}$ since $g(x) \in \mathfrak{o}[x]$. Using these linear combinations we can reduce the expression displayed above to a linear combination of $1, b, \ldots, b^{n-1}$ whose coefficients are in o since they are obtained by adding, subtracting, and multiplying the c_{ij} and the coefficients of $f(x)$ and $g(x)$, all of which are in o. Therefore every element of $\mathfrak{O}$ can be written in the form

$$\sum_{i=0}^{n-1} d_i b^i, \qquad d_i \in \mathfrak{o},$$

and so if $1, b, \ldots, b^{n-1}$ form a basis of K then it is a minimal basis. However, $1, b, \ldots, b^{n-1}$ span K over k, and so they must form a basis of K over k since the dimension of K over k is n. ‖

THEOREM 13. *Let $\bar{K}/\bar{k}$ be separable, $[K:k] = n$, and let a be an element of K such that $1, a, \ldots, a^{n-1}$ form a minimal basis of K/k. Let $f(x) = \mathrm{Irr}\ (k,a)$. Then $\mathfrak{D}_{K/k} = f'(a)\mathfrak{O}$.*

Proof. By Exercise 26,

$$f'(a)\mathfrak{O}' = \mathfrak{o} + a\mathfrak{o} + \cdots + a^{n-1}\mathfrak{o} = \mathfrak{O},$$

hence the result of the theorem. ‖

Now we shall describe all of the finite separable extensions K of k with $\bar{K}/\bar{k}$ separable. A polynomial $f(x) \in k[x]$ which has the form

$$f(x) = a_m x^m + a_{m-1}\pi^{r_{m-1}}x^{m-1} + \cdots a_1\pi^{r_1}x + a_0\pi,$$

where $a_0, a_1, \ldots, a_m$ are units in o and $r_i \geqslant 1$ for $i = 1, \ldots, m-1$, is called an *Eisenstein polynomial*. By the Eisenstein irreducibility criterion (Theorem 6) such a polynomial is irreducible in $k[x]$. If the extension K of k is obtained by adjoining to k a root of an Eisenstein polynomial in $k[x]$ then K is called an *Eisenstein extension* of k.

THEOREM 14. *Let K be a finite extension of k with $\bar{K}/\bar{k}$ separable. Then K is obtained from k by an unramified extension followed by an Eisenstein extension. Conversely, for any extension of k obtained in this manner, $\bar{K}/\bar{k}$ is separable.*

Proof. Let T be the inertia field of K. We shall show that K is an Eisenstein extension of T if and only if $\bar{K} = \bar{T}$. As usual we denote the ramification of K/k (and so of K/T) by e.

Suppose that $\bar{K} = \bar{T}$. As we have seen in the proof of Theorem 12, in this case $1, \Pi, \ldots, \Pi^{e-1}$ is a basis of K/T: let $f(x) = \mathrm{Irr}\ (T,\Pi)$. Every root of $f(x)$ in an algebraic closure C of k which contains K has the same valuation and so these roots are of the form $u\Pi$ where u is a unit of C. Thus every coefficient of $f(x)$ lies in $\mathfrak{O}_T$. Moreover, the constant term of $f(x)$ is $\pm N_{K/T}(\Pi)$ and $|N_{K/T}(\Pi)| = |\Pi|^e = |\pi|$, that is, the constant term of $f(x)$ has the form $v\pi$ where v is a unit in $\mathfrak{O}_T$. The only coefficient of $f(x)$ which is a unit in $\mathfrak{O}_T$ is the leading coefficient. Therefore $f(x)$ is an Eisenstein polynomial in $T[x]$.

Now suppose that K is obtained by adjoining to T a root of an Eisenstein polynomial $x^r + b_{r-1}x^{r-1} + \cdots + b_0$ in $T[x]$. Then $b_0, b_1, \ldots, b_{r-1} \in \mathfrak{P}_T$ and

$$a^r + b_{r-1}a^{r-1} + \cdots + b_1 a + b_0 = 0$$

where $K = k(a)$. From this we see that $|a^r| < 1$ and so $|a| < 1$. Then $|b_i a^i| < |b_0| = |\pi|$ for $i = 1, \ldots, r$ (with $b_r = 1$) and so $|a|^r = |\pi|$. Now we have $a = u\Pi^m$ for some integer m and some unit u in $\mathfrak{O}_K$ and $|\Pi|^e = |\pi|$. Hence $|\Pi|^e = |a|^r = |\Pi|^{rm}$ and so $r = [K:T] \leqslant e$. But e divides $[K:T]$ and therefore we have $e = r = [K:T]$. This implies that $f(K/T) = 1$, that is, $K = T$. $\|$

COROLLARY. *If K is a tamely ramified extension of k then*

$$\mathfrak{D}_{K/k} = \Pi^{e-1}\mathfrak{O}.$$

Proof. Let T be the inertia field of K. By the corollary to Theorem 10, $\mathfrak{D}_{K/k} = \mathfrak{D}_{K/T}$, and by Theorems 13 and 14,

$$\mathfrak{D}_{K/k} = f'(\Pi)\mathfrak{O},$$

where

$$f(x) = \mathrm{Irr}\,(T,\Pi) = x^e + b_{e-1}x^{e-1} + \cdots + b_1 x + b_0$$

is an Eisenstein polynomial in $T[x]$. Then

$$f'(\Pi) = e\Pi^{e-1} + (e-1)b_{e-1}\Pi^{e-2} + \cdots + b_1.$$

Hence $\mathfrak{D}_{K/T} = \Pi^n\mathfrak{O}$ where

$$n = \min \mathrm{ord}_K\,(e\Pi^{e-1}, (e-1)b_{e-1}\Pi^{e-2}, \ldots, b_1),$$

if one of the elements in the parentheses has ordinal strictly less than the ordinals of the others. We shall show that this is the case. For $i = 1, \ldots, e - 1$, b_i is divisible by π and so by Π^e which implies that $\mathrm{ord}_K b_i\Pi^{i-1} \geqslant e$. Since K/k is tamely ramified e is not divisible by char $\bar{k}$, which implies that e, considered as an element of K, does not belong to $\mathfrak{P}$. Therefore $\mathrm{ord}_K e\Pi^{e-1} = e - 1$, and we have $n = e - 1$. $\|$

5. Ramification groups

Throughout this section we shall assume that k is a local field whose residue class field $\bar{k}$ is perfect. Let K be a finite extension of k. We continue to use the notation of Section 1. If $\mathfrak{A}$ is an ideal of $\mathfrak{O}$ and if $a, b \in \mathfrak{O}$ we shall write $a \equiv b(\mathrm{mod}\ \mathfrak{A})$ whenever $a - b \in \mathfrak{A}$.

Assume that K is a Galois extension of k and let $G = G(K/k)$. For $i = 0, 1, \ldots,$ set

$$W_i = W_i(K/k) = \{\sigma \in G \mid \sigma(a) \equiv a(\mathrm{mod}\ \mathfrak{P}^{i+1})\ \text{for all}\ a \in \mathfrak{O}\}.$$

PROPOSITION 1. *For each i, W_i is a normal subgroup of G and $W_0 \supseteq W_1 \supseteq W_2 \supseteq \cdots$.*

Proof. The last part of the statement of the proposition follows immediately from the definition of the W_i. If $\sigma, \tau \in W_i$ then for all $a \in \mathfrak{O}$, $\sigma\tau(a) - a = \sigma\tau(a) - \sigma(a) + \sigma(a) - a = \sigma(\tau(a) - a) + (\sigma(a) - a) \in \mathfrak{P}^{i+1}$ and $\sigma^{-1}(a) - a = \sigma^{-1}(a - \sigma(a)) \in \mathfrak{P}^{i+1}$. Hence W_i is a subgroup of G. If $\sigma \in W_i$ and $\rho \in G$ then $\sigma(a) - a \in \mathfrak{P}^{i+1}$ and so $\sigma\rho(a) - \rho(a) \in \mathfrak{P}^{i+1}$. But then $\rho^{-1}\sigma\rho(a) - a = \rho^{-1}(\sigma\rho(a) - \rho(a)) \in \mathfrak{P}^{i+1}$. Hence $\rho^{-1}\sigma\rho \in W_i$. ‖

DEFINITION. The subgroup W_0 of G is called the *inertia group* of K/k and the subgroup W_i of G is called the *ith ramification group of K/k*.

PROPOSITION 2. *There is an integer i_0 such that $W_i = 1$ for $i \geqslant i_0$.*

Proof. By Theorem 12 there is an element $a \in \mathfrak{O}$ such that $1, a, \ldots, a^{n-1}$ is a minimal basis of K/k. Then $\sigma \in W_i$ if and only if $\sigma(a) \equiv a \pmod{\mathfrak{P}^{i+1}}$. Since $\bigcap_{i=0}^{\infty} \mathfrak{P}^{i+1} = 0$ we have $W_i = 1$ for all $i \geqslant \max_{\sigma \in G, \, \sigma \neq 1} \text{ord}(\sigma(a) - a)$. ‖

If $\sigma \in G(K/k)$ we can define an element $\bar{\sigma} \in G(\bar{K}/\bar{k})$ as follows: if $a \in \mathfrak{O}$ we set $\bar{\sigma}(\bar{a}) = \overline{\sigma(a)}$. This $\bar{\sigma}$ is well-defined, for if $\bar{a} = \bar{b}$ then $a - b \in \mathfrak{P}$ and therefore $\sigma(a) - \sigma(b) \in \mathfrak{P}$, or $\overline{\sigma(a)} = \overline{\sigma(b)}$. The mapping $\sigma \to \bar{\sigma}$ is clearly a homomorphism from $G(K/k)$ into $G(\bar{K}/\bar{k})$.

PROPOSITION 3. *The field $\bar{K}$ is a Galois extension of $\bar{k}$ and the homomorphism from $G(K/k)$ into $G(\bar{K}/\bar{k})$, described above, is onto.*

Proof. Since $\bar{K}/\bar{k}$ is separable there is an element $\bar{a} \in \bar{K}$ such that $\bar{K} = \bar{k}(\bar{a})$: we let $f(x)$ be a monic polynomial in $\mathfrak{o}[x]$ such that $\bar{f}(x) = \text{Irr}(\bar{k}, \bar{a})$. Then $f(x)$ is irreducible in $k[x]$. Let $\bar{f}(x) = (x - \bar{a})g(x)$ where $g(x) \in \bar{K}[x]$ and $g(\bar{a}) \neq 0$. It follows from Hensel's lemma, applied to K, that in $K[x]$ we have $f(x) = (x - b)h(x)$ where $\bar{b} = \bar{a}$. Thus $f(x)$ has one root in K and so splits in $K[x]$. But then $\bar{f}(x)$ splits in $\bar{K}[x]$ and so $\bar{K}$ is a splitting field over $\bar{k}$: as such it is normal over $\bar{k}$. Let $f(x) = (x - a_1) \cdots (x - a_f)$ in $K[x]$ so that $\bar{f}(x) = (x - \bar{a}_1) \cdots (x - \bar{a}_f)$ in $\bar{K}[x]$. An arbitrary element $\tau \in G(\bar{K}/\bar{k})$ is completely determined by the fact that, say, $\tau(\bar{a}_1) = \bar{a}_i$. There is an element $\sigma \in G(K/k)$ such that $\sigma(a_1) = a_i$. Then $\sigma \to \bar{\sigma} = \tau$ and so the homomorphism in question is onto. ‖

THEOREM 15. *If T is the inertia field of K then $W_0 = G(K/T)$.*

Proof. Let H be the kernel of the homomorphism from $G(K/k)$ onto $G(\bar{K}/\bar{k})$ which we have discussed above. Then $\sigma \in H$ if and only if $\bar{\sigma}(\bar{a}) = \bar{a}$ for all $a \in \mathfrak{O}$, that is, if and only if $\sigma(a) - a \in \mathfrak{P}$ for all $a \in \mathfrak{O}$. Thus $H = W_0$. If $\sigma \in G(K/T)$ then σ leaves fixed each element of T and $\bar{\sigma}$ leaves fixed each element of $T = \bar{K}$, that is, $\bar{\sigma}$ is the identity element of $G(\bar{K}/\bar{k})$. Thus $G(K/T) \subseteq W_0$. Then $f = \#G(\bar{K}/\bar{k}) = \#G(K/k)/\#W_0 \leqslant \#G(K/k)/\#G(K/T) = [T:k] = f$ and therefore $\#G(K/T) = \#W_0$ which implies that $G(K/T) = W_0$. ‖

COROLLARY.   *The inertia field of K is a Galois extension of k and $G(T/k) \cong G(\bar{K}/\bar{k})$.*

Let Π be a prime element of K and let e be the ramification of K/k. As we have seen before the elements $1, \Pi, \ldots, \Pi^{e-1}$ form a minimal basis of K/T. Thus for $i \geqslant 1$, $\sigma \in W_i$ if and only if $\sigma \in W_0$ and

$$\sigma(\Pi) \equiv \Pi(\text{mod } \mathfrak{P}^{i+1}).$$

Each element $\sigma \in W_0$ is completely determined by what it does to Π. Since $\sigma(\Pi)$ is also a prime element of K we have $\sigma(\Pi) = u_\sigma\Pi$ where u_σ is a unit in $\mathfrak{O}$. Consider the mapping ϕ from W_0 into $\bar{K}^*$ defined by

$$\phi(\sigma) = \bar{u}_\sigma.$$

If τ is another element of W_0 then $\tau(\Pi) = u_\tau\Pi$ where u_τ is a unit in $\mathfrak{O}$ and $\sigma\tau(\Pi) = \sigma(u_\tau\Pi) = \sigma(u_\tau)\sigma(\Pi) = u_\sigma\sigma(u_\tau)\Pi$ so that $u_{\sigma\tau} = u_\sigma\sigma(u_\tau)$. Since $\sigma \in W_0$, $\sigma(u_\tau) \equiv u_\tau(\text{mod } \mathfrak{P})$ and so $\overline{\sigma(u_\tau)} = \bar{u}_\tau$. Therefore, $\phi(\sigma\tau) = \bar{u}_\sigma\bar{u}_\tau = \phi(\sigma)\phi(\tau)$, so that ϕ is a homomorphism from W_0 into $\bar{K}^*$. We have σ in the kernel of ϕ if and only if $u_\sigma \equiv 1(\text{mod } \mathfrak{P})$, that is, if and only if $\sigma(\Pi) = u_\sigma\Pi \equiv \Pi(\text{mod } \mathfrak{P}^2)$. Thus the kernel of ϕ is W_1. This proves the first part of

THEOREM 16.   *The factor group W_0/W_1 is isomorphic to a finite subgroup of $\bar{K}^*$, it is cyclic, and has order prime to char $\bar{k}$.*

The last part of the statement of Theorem 16 is a consequence of the fact that every finite subgroup of the multiplicative group of a field is cyclic of order prime to the characteristic of the field. This is proved in the same manner as Theorem 23 in Chapter 1.

THEOREM 17.   *If $i \geqslant 1$ then the factor group W_i/W_{i+1} is isomorphic to a subgroup of $\bar{K}^+$. Hence if char $\bar{k} = 0$ we have $W_i = 1$ for $i \geqslant 1$, while if char $\bar{k} = p$ and $W_{i+1} \neq W_i$ then W_i/W_{i+1} is a direct product of cyclic groups of order p.*

Proof.   If $\sigma \in W_i$ we have, as above, $\sigma(\Pi) = u_\sigma\Pi$ where u_σ is a unit is $\mathfrak{O}$. Since $\sigma(\Pi) \equiv \Pi(\text{mod } \mathfrak{P}^{i+1})$ it follows that $u_\sigma \equiv 1(\text{mod } \mathfrak{P}^i)$. Thus $u_\sigma = 1 + a_\sigma\Pi^i$ where $a_\sigma \in \mathfrak{O}$. If $\tau \in W_i$ and $u_\tau = 1 + a_\tau\Pi^i$ then

$$u_{\sigma\tau} = u_\sigma\sigma(u_\tau) = (1 + a_\sigma\Pi^i)(1 + \sigma(a_\tau\Pi^i))$$
$$= 1 + (a_\sigma + a_\tau + (\sigma(a_\tau\Pi^i) - a_\tau\Pi^i)\Pi^{-i} + a_\sigma\sigma(a_\tau\Pi^i))\Pi^i.$$

Since $\sigma \in W_i$, $(\sigma(a_\tau\Pi^i) - a_\tau\Pi^i)\Pi^{-i} \in \mathfrak{P}$ and, as a result, if we define a mapping ψ from W_i into $\bar{K}^+$ by $\psi(\sigma) = \bar{a}_\sigma$ we have $\psi(\sigma\tau) = \bar{a}_\sigma + \bar{a}_\tau = \psi(\sigma) + \psi(\tau)$, that is, ψ is a homomorphism. An element $\sigma \in W_i$ is in the kernel of

ψ if and only if $a_\sigma \in \mathfrak{P}$, that is, if and only if $u_\sigma \equiv 1 \pmod{\mathfrak{P}^{i+1}}$, that is, if and only if $\sigma(\Pi) \equiv \Pi \pmod{\mathfrak{P}^{i+2}}$. Thus the kernel of ψ is W_{i+1} and therefore W_i/W_{i+1} is isomorphic to a finite subgroup of $\bar{K}^+$. If char $\bar{k} = 0$, the only such subgroup is the zero subgroup. If char $\bar{k} = p$ every nonzero element of $\bar{K}^+$ has additive order p, and every finite Abelian group in which every nonzero element has order p is a direct product of cyclic groups of order p. ‖

COROLLARY. *If V is the ramification field of K then $W_1 = G(K/V)$.*

Proof. Let F be the fixed field of W_1. The Galois group of F/T is isomorphic to W_0/W_1 and so by Theorem 16, $[F:T]$ is prime to char $\bar{k}$. In particular, $e(F/T)$ is prime to char $\bar{k}$, and we have $\bar{F} = \bar{T} = \bar{K}$, which is separable over $\bar{k}$. Thus F is a tamely ramified extension of k and so $F \subseteq V$. By Theorem 17, $[K:F]$ is either 1 or a power of char $\bar{k}$. But $[V:F] = e(V/F)$ is either 1 or is prime to char $\bar{k}$, and so we conclude that $F = V$. ‖

For further work it will be convenient for us to have a one-parameter family of subgroups of G, $\{W_t \mid t \geqslant 0\}$, which we shall now proceed to define. If t is a non-negative real number we set

$$W_t = \{\sigma \in W_0 \mid \operatorname{ord}_K(\sigma(\Pi) - \Pi) \geqslant t + 1\}.$$

If $\sigma, \tau \in W_t$ then

$$\begin{aligned}
\operatorname{ord}_K(\sigma^{-1}\tau(\Pi) - \Pi) &= \operatorname{ord}_K(\sigma^{-1}(\tau(\Pi) - \Pi) - \sigma^{-1}(\sigma(\Pi) - \Pi)) \\
&\geqslant \min(\operatorname{ord}_K(\tau(\Pi) - \Pi), \operatorname{ord}_K(\sigma(\Pi) - \Pi)) \\
&\geqslant t + 1,
\end{aligned}$$

so that $\sigma^{-1}\tau \in W_t$. Thus W_t is a subgroup of G. In fact,

$$W_t = W_i \qquad \text{where } i \text{ is the integer such that}$$
$$i - 1 < t \leqslant i.$$

The *Hasse function* of K/k is defined by

$$H_{K/k}(t) = \frac{1}{\#W_0} \int_0^t \#W_s \, ds.$$

Over any finite interval of the t-axis, $\#W_t$ has only a finite number of discontinuities and therefore the integral exists for all non-negative values of t.

PROPOSITION 4. *The function $H_{K/k}$ is continuous for $t > 0$, $H_{K/k}(0) = 0$, and $H_{K/k}$ is a strictly increasing function of t.*

PROPOSITION 5. *For each value of $t > 0$, $H_{K/k}$ has both left- and right-hand derivatives. If t is not an integer then*

$$D^- H_{K/k}(t) = D^+ H_{K/k}(t) = \#W_t/\#W_0,$$

while if t is an integer then

$$D^-H_{K/k}(t) = \#W_t/\#W_0, \qquad D^+H_{K/k}(t) = \#W_{t+1}/\#W_0.$$

PROPOSITION 6.　*The function $H_{K/k}$ is concave.*[1]

It follows from Proposition 4 that the Hasse function $H_{K/k}$ has an inverse function $h_{K/k}$. Then Proposition 5 holds with $H_{K/k}$ replaced by $h_{K/k}$, and $h_{K/k}$ is convex.

PROPOSITION 7.　*If t is an integer then $h_{K/k}(t)$ is an integer.*

Proof.　We shall show that if t is not an integer then $H_{K/k}(t)$ is not an integer. Suppose $t = [t] + \theta$ where $0 < \theta < 1$. Then

$$\frac{\#W_0}{\#W_t} H_{K/k}(t) = \frac{1}{\#W_t} \int_0^{[t]} \#W_s \, ds + \frac{1}{\#W_t} \int_{[t]}^{[t]+\theta} \#W_s \, ds$$

$$= \frac{\#W_1}{\#W_t} + \frac{\#W_2}{\#W_t} + \cdots + \frac{\#W_{[t]}}{\#W_t} + \frac{\#W_{[t]+1}}{\#W_t} \theta.$$

The last term in this sum is actually θ since $W_{[t]+1} = W_t$. Thus the right-hand side is not an integer and the same is true of the left-hand side. But $\#W_0/\#W_t$ is an integer and therefore $H_{K/k}(t)$ is not an integer. ‖

We shall now consider a field between k and K, say F, and determine the ramification groups of K/F and of F/k when F is a normal extension of k. The easier of these problems is the first and it is left as an exercise to the reader to show that for all $t \geqslant 0$,

$$W_t(K/F) = W_t(K/k) \cap G(K/F).$$

Assume that F/k is normal and let $G = G(K/k)$ and $H = G(K/F)$. By Theorem 12 there is an element $a \in \mathfrak{D}_K$ such that powers of a form a minimal basis of K/k and an element $b \in \mathfrak{D}_F$ such that powers of b form a minimal basis of F/k.

LEMMA 1.　*For all $\sigma \in G$ we have*

$$(\sigma(b) - b)\mathfrak{D}_K = \prod_{\eta \in H}(\sigma\eta(a) - a)\mathfrak{D}_K.$$

Proof.　The coefficients of

$$f(x) = \text{Irr}\,(F,a) = \prod_{\eta \in H}(x - \eta(a))$$

---

[1] For the properties of concave and convex functions, see E. F. Beckenbach, "Convex functions," *Bulletin of the American Mathematical Society, 54*, 1948, pp. 439–60.

are in $\mathfrak{D}_F$ and

$$(\sigma f)(x) = \text{Irr}\,(\sigma(F),\sigma(a)) = \prod_{\eta \in H}(x - \sigma\eta(a)).$$

If $c \in \mathfrak{D}_F$ then $c = c_r b^r + \cdots + c_1 b + c_0$, where $c_0, c_1, \ldots, c_r \in \mathfrak{o}$, and $\sigma(c) - c = c_r(\sigma(b)^r - b^r) + \cdots + c_1(\sigma(b) - b) = (\sigma(b) - b)d$ where $d \in \mathfrak{D}_F$. Hence every coefficient of $(\sigma f)(x) - f(x)$ is in $(\sigma(b) - b)\mathfrak{D}_K$. Therefore,

$$\prod_{\eta \in H}(\sigma\eta(a) - a)\mathfrak{D}_K = ((\sigma f)(a) - f(a))\mathfrak{D}_K \subseteq (\sigma(b) - b)\mathfrak{D}_K.$$

On the other hand, let $g(x) \in \mathfrak{o}[x]$ be such that $b = g(a)$ and note that $(\sigma g)(x) = g(x)$. The polynomial $g(x) - b$ is in $F[x]$ and has a as a root: hence it is divisible by $f(x)$ in $F[x]$ and we let $g(x) - b = f(x)h(x)$ where $h(x) \in \mathfrak{D}_F[x]$. Then $\sigma(b) - b = \sigma(b) - g(a) = \sigma(b) - (\sigma g)(a) = -(\sigma f)(a)(\sigma h)(a) = -((\sigma f)(a) - f(a))(\sigma h)(a)$, since $f(a) = 0$, and so

$$(\sigma(b) - b)\mathfrak{D}_K \subseteq ((\sigma f)(a) - f(a))\mathfrak{D}_K. \;\|$$

For $\sigma \in W_0$, $\sigma \neq 1$, we define $w_{K/k}(\sigma)$ to be the integer such that

$$\sigma \in W_{w_{K/k}(\sigma)}(K/k) \quad \text{but} \quad \sigma \notin W_{w_{K/k}(\sigma)+1}(K/k).$$

Further we set $w_{K/k}(1) = \infty$. If ρ_S is the characteristic function of the subset S of G, that is, if

$$\rho_S(\sigma) = \begin{cases} 1 & \text{if} \quad \sigma \in S, \\ 0 & \text{if} \quad \sigma \notin S, \end{cases}$$

then

$$w_{K/k}(\sigma) + 1 = \sum_{i=0}^{\infty} \rho_{W_i(K/k)}(\sigma).$$

If $\sigma \neq 1$ we also have

$$w_{K/k}(\sigma) + 1 = \text{ord}_K\,(\sigma(a) - a).$$

Let $\bar{\sigma} = \sigma H$ be a typical nonidentity element of G/H, which we may regard as the Galois group of F/k. We can define $w_{F/k}(\bar{\sigma}) + 1 = \text{ord}_F\,(\sigma(b) - b)$.

LEMMA 2. *If $\sigma H \cap W_i(K/k)$ is not empty for some i then*

$$w_{F/k}(\bar{\sigma}) + 1 = \frac{1}{\#W_0(K/F)} \sum_{i=0}^{m(\sigma)} \#W_i(K/F)$$

$$= H_{K/F}(m(\sigma)) + 1,$$

where $m(\sigma)$ is the integer defined by $\sigma \in W_{m(\sigma)}(K/k)H$ but $\sigma \notin W_{m(\sigma)+1}(K/k)H$.

Proof. Since $\#W_0(K/F)$ is the ramification of K/F we have, making use of Lemma 1,

$$
\begin{aligned}
\#W_0(K/F)(w_{F/k}(\bar{\sigma}) + 1) &= \#W_0(K/F)\,\mathrm{ord}_F\,(\sigma(b) - b) \\
&= \mathrm{ord}_K\,(\sigma(b) - b) = \mathrm{ord}_K \prod_{\eta \in H}(\sigma\eta(a) - a) \\
&= \sum_{\eta \in H} \mathrm{ord}_K\,(\sigma\eta(a) - a) = \sum_{\eta \in H}(w_{K/k}(\sigma\eta) + 1) \\
&= \sum_{\eta \in H} \sum_{i=0}^{\infty} \rho_{W_i(K/k)}(\sigma\eta) \\
&= \sum_{i=0}^{\infty} \sum_{\eta \in H} \rho_{W_i(K/k)}(\sigma\eta) \\
&= \sum_{i=0}^{\infty} \#\,(\sigma H \cap W_i(K/k)).
\end{aligned}
$$

Assume that $\sigma H \cap W_i(K/k)$ is not empty: let $\sigma\omega \in W_i(K/k)$ where $\omega \in H$. Then $\sigma\eta \in W_i(K/k)$ if and only if $\sigma\omega\omega^{-1}\eta \in W_i(K/k)$, that is, if and only if $\omega^{-1}\eta \in (\sigma\omega)^{-1}W_i(K/k)$, that is, if and only if $\omega^{-1}\eta \in H \cap W_i(K/k)$. Thus we have a one-one correspondence between the elements of $\sigma H \cap W_i(K/k)$ and the elements of $H \cap W_i(K/k)$. Therefore (see Exercise 37)

$$
\begin{aligned}
\#W_0(K/F)(w_{F/k}(\bar{\sigma}) + 1) &= \sum_{i=0}^{m(\sigma)} \#(H \cap W_i(K/k)) \\
&= \sum_{i=0}^{m(\sigma)} \#W_i(K/F). \;\|
\end{aligned}
$$

THEOREM 18. *The ramification groups of F/k are given by*

$$W_t(F/k) = W_{h_{K/F}(t)}(K/k)H/H.$$

Proof. Since $h_{K/F}$ is the inverse function of $H_{K/F}$ we wish to prove that

$$W_{H_{K/F}(t)}(F/k) = W_t(K/k)H/H.$$

Let $\bar{\sigma} = \sigma H \in G/H$, $\sigma H \neq H$. Then $\bar{\sigma} \in W_{H_{K/F}(t)}(F/k)$ if and only if $w_{F/k}(\bar{\sigma}) \geqslant H_{K/F}(t)$. This is equivalent to $H_{K/F}(m(\sigma)) \geqslant H_{K/F}(t)$, or $m(\sigma) \geqslant t$, or $\sigma \in W_t(K/k)H$, or $\bar{\sigma} \in W_t(K/k)H/H$. $\|$

THEOREM 19. *For $t \geqslant 0$ we have*

$$H_{K/k}(t) = H_{F/k}(H_{K/F}(t))$$

so that

$$h_{K/k}(t) = h_{K/F}(h_{F/k}(t)).$$

Proof. For $t \geqslant 0$,

$$H_{F/k}(t) = \frac{1}{\#W_0(F/k)} \int_0^t \#W_s(F/k) \, ds,$$

$$= \frac{1}{\#(W_0(K/k)H/H)} \int_0^t \# \left(\frac{W_{h_{K/F}(s)}(K/k)H}{H} \right) \, ds,$$

$$= \frac{\#(W_0(K/k) \cap H)}{\#W_0(K/k)} \int_0^t \frac{\#W_{h_{K/F}(s)}(K/k)}{\#(W_{h_{K/F}(s)}(K/k) \cap H)} \, ds,$$

$$= \frac{\#W_0(K/F)}{\#W_0(K/k)} \int_0^t \frac{\#W_{h_{K/F}(s)}(K/k)}{\#W_{h_{K/F}(s)}(K/F)} \, ds.$$

Now make a change of variable in the integral, replacing s by $H_{K/F}(s)$. Then

$$H_{F/k}(t) = \frac{\#W_0(K/F)}{\#W_0(K/k)} \int_0^{h_{K/F}(t)} \frac{\#W_s(K/k)}{\#W_s(K/F)} \, dH_{K/F}(s).$$

At all but a finite number of points of the interval of integration we have, by Proposition 5,

$$dH_{K/F}(s) = \frac{\#W_s(K/F)}{\#W_0(K/F)} \, ds,$$

so that

$$H_{F/k}(t) = \frac{\#W_0(K/F)}{\#W_0(K/k)} \int_0^{h_{K/F}(t)} \frac{\#W_s(K/k)}{\#W_s(K/F)} \frac{\#W_s(K/F)}{\#W_0(K/F)} \, ds$$

$$= \frac{1}{\#W_0(K/k)} \int_0^{h_{K/F}(t)} \#W_s(K/k) \, ds$$

$$= H_{K/k}(h_{K/F}(t)).$$

If we replace t by $H_{K/F}(t)$ we get $H_{K/k}(t) = H_{F/k}(H_{K/F}(t))$. ∥

EXERCISES

Section 1

1. Let k be a non-Archimedean valued field and let L and K be finite extensions of k with $k \subseteq L \subseteq K$. Show that $e(K/k) = e(K/L)e(L/k)$.

2. Let k be a non-Archimedean valued field and let K be a finite extension of k. Show that $ef = n$ if and only if there is a basis $a_1, \ldots, a_n$ of K over k such that for all $c_1, \ldots, c_n \in k$,

$$\left| \sum_{i=1}^n c_i a_i \right| = \max_{1 \leqslant i \leqslant n} |c_i a_i|.$$

3. Let k be a local field and let π be a prime element of k. If $a \in k^*$ then, by the lemma preceding the proof of Theorem 2, if $\mathfrak{R}$ is a complete set of representatives in $\mathfrak{o}$ of the elements of $\bar{k}$, we can write uniquely

$$a = \sum_{i=n}^{\infty} c_i \pi^i, \quad n = \operatorname{ord} a, \quad c_i \in \mathfrak{R}, \quad c_n \notin \mathfrak{p};$$

we call this the *expansion* of a with respect to π and $\mathfrak{R}$.

  (a) Show that in Q_p the set $\mathfrak{R} = \{0, 1, \ldots, p-1\}$ is a complete set of representatives in $\mathfrak{o}$ of the elements of $\bar{Q}_p$.

  (b) Obtain the expansions of ± 3, $\pm\frac{1}{3}$, and ± 5 in Q_2 with respect to 2 and $\mathfrak{R}$.

  (c) Obtain the expansions of -1, -2, $\pm\frac{1}{2}$, and $\pm\frac{1}{5}$ in Q_3 with respect to 3 and $\mathfrak{R}$.

  (d) Show that the expansion of a in Q_p with respect to p and $\mathfrak{R}$ is periodic if and only if $a \in Q$.

4. Let k be a discrete non-Archimedean valuated field and let K be a finite extension of k. Show that for all $a \in k$, $\operatorname{ord}_K a = e(K/k)\operatorname{ord}_k a$.

5. Let k be a local field and K a finite extension of k. Let $u_1, \ldots, u_f$ be as in the proof of Theorem 2. Show that the basis of K/k made up of the elements $u_i \Pi^j$, $i = 1, \ldots, f$, $j = 0, \ldots, e-1$, is a minimal basis, that is, it has the property that a linear combination of these elements with coefficients in k lies in $\mathfrak{D}$ if and only if the coefficients are all in $\mathfrak{o}$.

6. Let $k = Q_2(\sqrt{2}, \sqrt[4]{2}, \sqrt[8]{2}, \ldots)$ and let $K = k(i)$ where $i^2 = -1$. Show that $e = f = 1$ but $n = 2$, so that $ef < n$. Show that k is not complete, so that we have another example of a relatively complete, but not complete, valuated field.

7. Let k be a relatively complete non-Archimedean valuated field which is discrete. Let K be an infinite algebraic extension of k. Investigate whether or not the prolongation of the valuation on k to a valuation on K is discrete.

8. Let k be a complete non-Archimedean valuated field and let K be a finite extension of k. Show that if a prime q divides ef then q divides n.

9. With k and K as in Exercise 8, let char $\bar{k} = p$. Show that if $[K:k] = q$, where q is a prime and $q \neq p$, then $ef = q$. (This is not true if $q = p$, as we see from Exercise 6.)

10. With k and K as in Exercise 9, show that for some integer t, $n = efp^t$.

11. Let k be a local field and let π be a prime element of k. For $i = 0$, $1, \ldots,$ let $\mathfrak{R}_i$ be a complete set of representatives in $\mathfrak{o}$ of the elements of $\bar{k}$. Also, for each i, let $S_i = \{1 + c\pi^i \mid c \in \mathfrak{R}_i\}$. Show that each $a \in k^*$ can be written uniquely in the form

$$a = \pi^n b \prod_{i=1}^{\infty} s_i, \quad n = \operatorname{ord} a, \quad b \in \mathfrak{R}_0, \quad s_i \in S_i.$$

(Use the theory of infinite products which you developed in answer to Exercise 18 of Chapter 3.)

Section 2

12. Let k be a local field and $K = k(a)$ where a is a root of a monic polynomial $f(x) \in \mathfrak{o}[x]$ such that $\bar{f}(x)$ has distinct roots in an algebraic closure of $\bar{k}$. Show that K/k is unramified.

13. Let k, C, etc., be as in Section 2. Let K be an unramified extension of k in C and let L be a finite extension of k in C. Show that LK/L is unramified.

14. Same as Exercise 13 with "unramified" replaced by "tamely ramified."

15. Let k be a field and let $F = k\{t\}$ be the set of all formal power series

$$a = \sum a_n t^n, \quad a_n \in k \text{ and } a_n \neq 0$$

for only a finite number of negative n.
If $b = \Sigma b_n t^n$ we define $a + b$ and ab by $a + b = \Sigma (a_n + b_n)t^n$ and $ab = \Sigma c_n t^n$ where $c_n = \sum_{r+s=n} a_r b_s$.

  (a) Show that with respect to these operations F is a field which contains the field $k(t)$ of rational functions over k as a subfield.
  (b) If $a \in F$, $a \neq 0$, and if $a_r = 0$ for $r < n$ but $a_n \neq 0$, we set $|a| = e^{-n}$: also set $|0| = 0$. Show that $|\ |$ is a discrete non-Archimedean valuation on F and that t is a prime element of F.
  (c) Show that F is the completion of $k(t)$ with respect to $|\ |_t$.
  (d) Show that $\bar{F}$ is isomorphic to k and so may be identified with k.
  (e) Show that if K/k is separable then $K\{t\}$ is an unramified extension of F.
  (f) Show that if L is an unramified extension of F then there is a separable extension K of k such that $L = K\{t\}$.

16. Let k be a local field, and let K be a tamely ramified extension of k. Show that K is a separable extension of k.

17. Let k be a local field such that $\bar{k} = GF(p^r)$. Show that K/k is unramified if and only if K is a splitting field over k of the polynomial $x^{p^{nr}} - x$ where $n = [K:k]$.

18. Suppose $\bar{k}$ is finite, where k is a local field, and let K be an unramified extension of k. Then K/k is normal and separable (see Exercises 16 and 17). Let q be the number of elements in $\bar{k}$. Show that $G(K/k)$ is cyclic and is generated by the unique element σ such that $\sigma(a) - a^q \in \mathfrak{P}$ for all $a \in \mathfrak{O}$.

19. Let k and K be as in Exercise 18. Show that if u is a unit in k then there is a unit v in K such that $u = N_{K/k}(v)$.

20. Let k and K be as in Exercise 18. Show that $k^*/N_{K/k}(K^*)$ is cyclic of order $n = [K:k]$.

Sections 3 and 4

21. Let k be a local field, $\mathfrak{o}$ the ring of integers of k, $\mathfrak{p}$ the ideal of nonunits of $\mathfrak{o}$, and U the group of units of k. Considering $\mathfrak{p}$ and its powers as additive

groups, and with $n > 0$ and r integers, show that

(a) all powers of $\mathfrak{p}$ are isomorphic,

(b) $\mathfrak{p}^r/\mathfrak{p}^{r+n} \cong \mathfrak{o}/\mathfrak{p}^n$, and in particular, $\mathfrak{p}^r/\mathfrak{p}^{r+1} \cong \mathfrak{o}/\mathfrak{p} = \bar{k}^+$,

(c) if $\bar{k} = GF(p^s)$ then $(\mathfrak{p}^r : \mathfrak{p}^{r+1}) = p^s$.

22. Let k, $\mathfrak{o}$, etc., be as in Exercise 21 and for $r > 0$ set $U_r = 1 + \mathfrak{p}^r = \{1 + a \mid a \in \mathfrak{p}^r\}$, and $U_0 = U$. Show that

(a) U_r is a subgroup of U.

(b) $U_0/U_1 \cong \bar{k}^*$,

(c) for $r > 0$, $U_r/U_{r+1} \cong \bar{k}^+$.

23. Show that in a natural way we have $U \cong \varprojlim U/U_r$.

24. Let k be an arbitrary field and let K be a finite separable extension of k. Let $K = k(a)$ with Irr $(k,a) = f(x)$, $n = \deg f(x)$, and let $f(x) = (x - a_1) \cdots (x - a_n)$ over some algebraic closure of k which contains K. Show that

$$\sum_{j=1}^{n} \frac{f(x)}{x - a_j} \frac{a_j^{\,i}}{f'(a_j)} = x^i, \qquad i = 0, \quad 1, \ldots, \quad n - 1.$$

(Hint. Use the Lagrange interpolation formula.)

25. With the notation as in Exercise 24, let $f(x) = (x - a)(b_{n-1}x^{n-1} + \cdots + b_1 x + b_0)$ in $K[x]$. Show that

$$\frac{b_0}{f'(a)}, \qquad \frac{b_1}{f'(a)}, \qquad \cdots, \qquad \frac{b_{n\,1}}{f'(a)}$$

is the complementary basis of $1, a, \ldots, a^{n-1}$.

26. Let k be a local field and let K, a, etc., be as in Exercise 24. Show that if $1, a, \ldots, a^{n-1}$ is a minimal basis of K/k then

$$f'(a)\mathfrak{D}' = \mathfrak{o} + a\mathfrak{o} + \cdots + a^{n-1}\mathfrak{o}.$$

(Hint. Express each b_i in terms of powers of a.)

27. Let $K = Q_p(\sqrt{p})$. Show that $1, \sqrt{p}$ form a minimal basis of K/Q_p. If $\mathfrak{P}$ is the ideal of nonunits of the ring of integers of K show that $\mathfrak{D}_{K/Q_p} = \mathfrak{P}$ if $p \neq 2$ and $\mathfrak{P}^3$ if $p = 2$.

Section 5

In the Exercises for this section, k is a local field with $\bar{k}$ perfect and K is a finite Galois extension of k. All notation will be as in Section 5.

28. If $a \in K$ and if powers of a form a minimal basis of K/k, show that $\sigma \in W_i$ if and only if ord $(\sigma(a) - a) \geq i + 1$.

29. Let $\tau \in W_0$ and $\sigma \in W_i$ $(i > 0)$. Show that $\tau\sigma\tau^{-1}\sigma^{-1} \in W_{i+1}$ if and only if either $\tau^i \in W_1$ or $\sigma \in W_{i+1}$. Using this fact show that

(a) W_i/W_{i+1} is in the center of W_1/W_{i+1},

(b) if K/k is Abelian and if $e = dp^r$ where $p = $ char $\bar{k}$ and $p \nmid d$, and if $W_i \neq W_{i+1}$, then d divides i.

30. Let $\sigma \in W_i$ and $\tau \in W_j$ where $i, j \geqslant 1$. Show that $\sigma\tau\sigma^{-1}\tau^{-1} \in W_{i+j}$. Show that W_i/W_{2i} is Abelian.

31. Show that if char $\bar{k} = p$ and if $W_i \neq W_{i+1}$ and $W_j \neq W_{j+1}$ where $i, j \geqslant 1$ then $i \equiv j \pmod{p}$.

32. Show that if $\bar{k}$ is finite then K is a solvable extension of k.

33. Show that

$$\mathfrak{D}_{K/k} = \Pi^{\sum\limits_{i=0}^{\infty} (\#W_i - 1)} \mathfrak{D}.$$

34. Let $K = Q_p(\sqrt{p})$. Determine the ramification groups of K/Q_p and their fixed fields.

35. Give proofs of Propositions 4, 5, and 6 of Section 5.

36. If $k \subseteq F \subseteq K$ show that for all $t \geqslant 0$, $W_t(K/F) = W_t(K/k) \cap G(K/F)$.

37. Show that $m(\sigma)$, as defined in Lemma 2 of Section 5, is the greatest integer i such that $\sigma H \cap W_i(K/k)$ is not empty. (This fact is used in the last step of the proof of Lemma 2.)

38. Let k be a local field and K a finite separable extension of k (in this exercise we do not require that K/k be normal). Let L be a finite Galois extension of k such that $k \subseteq K \subseteq L$. For $t \geqslant 0$ set $H_{K/k}(t) = H_{L/k}(h_{L/K}(t))$ and $h_{K/k}(t) = H_{L/K}(h_{L/k}(t))$.

(a) Show that the functions $H_{K/k}$ and $h_{K/k}$ are independent of the choice of the Galois extension L of k, and that they are inverse functions of each other.

(b) Let $k \subseteq F \subseteq K$. Show that for all $t \geqslant 0$, $H_{K/k}(t) = H_{F/k}(H_{K/F}(t))$ and $h_{K/k}(t) = h_{K/F}(h_{F/k}(t))$.

39. For $t \geqslant 0$ set $W^t = W^t(K/k) = W_{h_{K/k}(t)}(K/k)$.

(a) Show that $W^0 = W_0$ and that $W^t = 1$ for t sufficiently large.

(b) Show that $h_{K/k}(t) = \int_0^t (W^0 : W^s)\,ds$.

(c) If H is a normal subgroup of G, and if F is the fixed field of H, show that

$$W^t(F/k) = W^t(K/k)H/H.$$

40. Let $[K:k]$ be a prime p and let σ be a generator of G. Assume that $\bar{K} = \bar{k}$. Let $r = \operatorname{ord}_K (\sigma(\Pi) - \Pi) - 1$. Show that

(a) $G = W_0 = \cdots = W_r$ and $W_{r+1} = 1$,

(b) $r \neq 0$ if and only if $p = $ char $\bar{k}$,

(c) $h_{K/k}(t) = \begin{cases} t & \text{if } t \leqslant r, \\ r + p(t - r) & \text{if } t \geqslant r, \end{cases}$

(d) $\mathfrak{D}_{K/k} = \mathfrak{P}^{(r+1)(p-1)}$.

CHAPTER 5

Dedekind Fields

1. The fundamental theorem of Dedekind fields

Let k be a field and let $\mathfrak{B}$ be a family of inequivalent, nontrivial, discrete non-Archimedean valuations on k. We let the elements of $\mathfrak{B}$ be indexed by a set I which may have any nonzero cardinality. For each $\alpha \in I$ let $\mathfrak{o}_\alpha$ be the ring of integers of k with respect to the valuation $|\ |_\alpha$ in $\mathfrak{B}$ and set

$$\mathfrak{o} = \bigcap_{\alpha \in I} \mathfrak{o}_\alpha.$$

This is a subring of k and for each $\alpha \in I$ we let $\mathfrak{p}_\alpha$ be the intersection of $\mathfrak{o}$ and the ideal of nonunits of $\mathfrak{o}_\alpha$. Then $\mathfrak{p}_\alpha$ is a prime ideal of the ring $\mathfrak{o}$. We shall show that when certain conditions are imposed on the set of valuations $\mathfrak{B}$, every nonzero ideal of $\mathfrak{o}$ can be written uniquely as a product of prime ideals of $\mathfrak{o}$. A commutative integral domain with unity which has this property is called a *Dedekind domain*. This is the reason for the name we have chosen for the type of system we shall study in this chapter.

DEFINITION. The composite object $(k, \mathfrak{B})$ will be called a *Dedekind field* if
(A) the ring $\mathfrak{o}$ has k as a field of quotients,
(B) for each nonzero $a \in \mathfrak{o}$ there are only a finite number of $\alpha \in I$ for which $|a|_\alpha \neq 1$,
(C) for $\alpha, \beta \in I$ with $\alpha \neq \beta$ there is an element $a \in \mathfrak{o}$ such that $|a|_\alpha < 1$ and $|a|_\beta = 1$, and
(D) $\mathfrak{o}$ is the only ideal of $\mathfrak{o}$ which contains a unit from each $\mathfrak{o}_\alpha$.

We note that if $(k, \mathfrak{B})$ satisfies (A)–(D) and if any valuation in $\mathfrak{B}$ is replaced by an equivalent valuation, then (A)–(D) will continue to hold.

If $(k, \mathfrak{B})$ is a Dedekind field then the ring $\mathfrak{o}$ is called its *ring of integers*.

If $\mathfrak{B}$ consists of a single valuation then $(k,\mathfrak{B})$ is a Dedekind field. The requirements (A) and (D) certainly hold, (B) holds since I is finite, and (C) holds vacuously. Our interest lies in the more significant examples in which $\mathfrak{B}$ contains more than one valuation and, in particular, in the example we shall now give. The importance of this example will become more apparent after the next section is read.

Consider the field Q of rational numbers and let

$$\mathfrak{B} = \{| \ |_p \, | \, p \text{ a prime}\}.$$

We shall show that $(Q,\mathfrak{B})$ is a Dedekind field. Clearly $J \subseteq \mathfrak{o}$. If $a \in Q$ is not zero and is reduced to lowest terms then $|a|_p > 1$ if and only if p divides the denominator of a. Hence, if $a \in \mathfrak{o}$, that is, $|a|_p \leqslant 1$ for all primes p, then $a \in J$. Therefore, $\mathfrak{o} = J$ and we see that Q is a field of quotients of $\mathfrak{o}$: thus (A) holds. Again, if a is reduced to lowest terms, then $|a|_p \neq 1$ if and only if p divides either the numerator or the denominator of a. Hence (B) holds. If p and q are distinct primes then $|p|_p < 1$ and $|p|_q = 1$ and therefore (C) holds. Finally, let $\mathfrak{a}$ be an ideal of J such that for each prime p there is an element $u_p \in \mathfrak{a}$ such that $|u_p|_p = 1$. The ideal $\mathfrak{a}$ is a principal ideal with a generator m. If $m \neq \pm 1$ then for every prime divisor p of m we must have $|a|_p < 1$ for every $a \in \mathfrak{a}$. Since this is not true for any p we have $m = \pm 1$ and so $\mathfrak{a} = J = \mathfrak{o}$: thus (D) holds.

Let $(k,\mathfrak{B})$ be a Dedekind field and let $\mathfrak{o}$ be its ring of integers. A nonempty subset A of k is called a *fractional ideal* of $(k,\mathfrak{B})$ if

    (i) $a - b \in A$ for all $a, b \in A$,

    (ii) $ca \in A$ for all $a \in A$ and $c \in \mathfrak{o}$, and

    (iii) there is a nonzero element $d \in \mathfrak{o}$ such that $dA \subseteq \mathfrak{o}$. (See Section 3 of Chapter 4, where we defined fractional ideal in the case of a local field.) If A and B are fractional ideals of $(k,\mathfrak{B})$ we define their product to be

$$AB = \{\text{finite sums } \textstyle\sum a_i b_i \, | \, a_i \in A, \, b_i \in B\}.$$

This is a fractional ideal of $(k,\mathfrak{B})$ and we see immediately that $AB = BA$ and that if C is a fractional ideal of $(k,\mathfrak{B})$ then $A(BC) = (AB)C$. If A is a fractional ideal of $(k,\mathfrak{B})$ and if $A \subseteq \mathfrak{o}$ then it follows from (i) and (ii) that A is an ideal of $\mathfrak{o}$. Conversely, every ideal of $\mathfrak{o}$ is a fractional ideal of $(k,\mathfrak{B})$.

Let $F(k,\mathfrak{B})$ be the set of all nonzero fractional ideals of $(k,\mathfrak{B})$. We shall show that $F(k,\mathfrak{B})$ is a group with respect to the binary operation defined in the preceding paragraph. It has been shown in Section 3 of Chapter 4 that this is indeed the case when k is a local field: in fact, it has been shown there that in this case the set of nonzero fractional ideals is a cyclic group with respect to this multiplication and is generated by the ideal of nonunits of the ring of integers of k. In that discussion we did not use the fact that k is complete.

If A and B are fractional ideals of $(k,\mathfrak{B})$ we set

$$A/B = \{c \in k \mid cB \subseteq A\}.$$

PROPOSITION 1. *If $B \neq 0$ then A/B is a fractional ideal of $(k,\mathfrak{B})$.*

Proof. Let $d \in \mathfrak{o}$, $d \neq 0$, be such that $dA \subseteq \mathfrak{o}$. Since $B \neq 0$ there is a nonzero element $b \in B$. If $c \in A/B$ then $cb \in A$ and so $cbd \in \mathfrak{o}$. Therefore $(bd)(A/B) \subseteq \mathfrak{o}$. The other requirements for a fractional ideal clearly hold for A/B, so that A/B is a fractional ideal. $\parallel$

If $A \in F(k,\mathfrak{B})$ then $\mathfrak{o}A \subseteq A$ and since $1 \in \mathfrak{o}$ we have $A \subseteq \mathfrak{o}A$. Hence $\mathfrak{o}A = A$ and so $\mathfrak{o}$ is a unity element of $F(k,\mathfrak{B})$. We now set $A^{-1} = \mathfrak{o}/A$. If we can show that $A^{-1}A = \mathfrak{o}$ then it will follow that $F(k,\mathfrak{B})$ is a group with respect to the multiplication we have defined. We have $A^{-1}A \subseteq \mathfrak{o}$ so that $A^{-1}A$ is an ideal of $\mathfrak{o}$. Because of (D) our task will be accomplished if we show that $A^{-1}A$ contains a unit in $\mathfrak{o}_\alpha$ for each $\alpha \in I$.

If $\alpha \in I$ we denote the ordinal function with respect to the valuation $\mid \; \mid_\alpha$ by $\operatorname{ord}_\alpha$. Recall from Section 1 of Chapter 4 that $\operatorname{ord}_\alpha$ is so defined that if π_α is a prime element of k with respect to $\mid \; \mid_\alpha$ then $\operatorname{ord}_\alpha \pi_\alpha = 1$. If $A \in F(k,\mathfrak{B})$ we define

$$\operatorname{ord}_\alpha A = \min_{a \in A} \operatorname{ord}_\alpha a.$$

We can now restate (D) in the following way: if $\mathfrak{a}$ is an ideal of $\mathfrak{o}$ such that $\operatorname{ord}_\alpha \mathfrak{a} = 0$ for all $\alpha \in I$ then $\mathfrak{a} = \mathfrak{o}$. We shall now show in a series of propositions that for each $A \in F(k,\mathfrak{B})$ we have $\operatorname{ord}_\alpha A^{-1}A = 0$ for all $\alpha \in I$.

PROPOSITION 2. *If $A, B \in F(k,\mathfrak{B})$ and $\alpha \in I$ then $\operatorname{ord}_\alpha AB = \operatorname{ord}_\alpha A + \operatorname{ord}_\alpha B$.*

Proof. Let $a \in A$ and $b \in B$ be such that $\operatorname{ord}_\alpha a = \operatorname{ord}_\alpha A$ and $\operatorname{ord}_\alpha b = \operatorname{ord}_\alpha B$. Then $\operatorname{ord}_\alpha AB \leqslant \operatorname{ord}_\alpha ab = \operatorname{ord}_\alpha a + \operatorname{ord}_\alpha b = \operatorname{ord}_\alpha A + \operatorname{ord}_\alpha B$. On the other hand, for certain elements $a_i \in A$ and $b_i \in B$ we have $\operatorname{ord}_\alpha AB = \operatorname{ord}_\alpha \Sigma \, a_i b_i \geqslant \min \operatorname{ord}_\alpha a_i b_i \geqslant \min \operatorname{ord}_\alpha a_i + \min \operatorname{ord}_\alpha b_i \geqslant \operatorname{ord}_\alpha A + \operatorname{ord}_\alpha B$. $\parallel$

PROPOSITION 3. *If $A \in F(k,\mathfrak{B})$ then $\operatorname{ord}_\alpha A \neq 0$ for only a finite number of $\alpha \in I$.*

Proof. Suppose that $A \subseteq \mathfrak{o}$ and let $a \in A$, $a \neq 0$. Then $0 \leqslant \operatorname{ord}_\alpha A \leqslant \operatorname{ord}_\alpha a$ for all $\alpha \in I$. However, by (B), $\operatorname{ord}_\alpha a \neq 0$ for only a finite number of $\alpha \in I$ and it is only for these α that we could possibly have $\operatorname{ord}_\alpha A \neq 0$. If $A \subseteq \mathfrak{o}$ let $d \in \mathfrak{o}$, $d \neq 0$, be such that $dA \subseteq \mathfrak{o}$. Then $\operatorname{ord}_\alpha A = \operatorname{ord}_\alpha dA - \operatorname{ord}_\alpha d$. Since $\operatorname{ord}_\alpha dA \neq 0$ or $\operatorname{ord}_\alpha d \neq 0$ for only a finite number of $\alpha \in I$ the same must be true of $\operatorname{ord}_\alpha A$. $\parallel$

PROPOSITION 4. *If $\alpha_1, \ldots, \alpha_n$ are distinct elements of I then there are elements $a, b \in \mathfrak{o}$ such that $|a|_{\alpha_1} < 1$, $|a|_{\alpha_i} = 1$ for $i = 2, \ldots, n$, and $|b|_{\alpha_1} = 1$, $|b|_{\alpha_i} < 1$ for $i = 2, \ldots, n$.*

Proof. Let $2 \leqslant i \leqslant n$. By (C) there is an element $b_i \in \mathfrak{o}$ such that $|b_i|_{\alpha_1} = 1$ and $|b_i|_{\alpha_i} < 1$. Let $b = b_2 \cdots b_n$. Then $|b|_{\alpha_1} = 1$ and $|b|_{\alpha_i} < 1$ for $i = 2, \ldots, n$. It follows from this that if $2 \leqslant j \leqslant n$ then there is an element $a_j \in \mathfrak{o}$ such that $|a_j|_{\alpha_j} = 1$ and $|a_j|_{\alpha_i} < 1$ for $i = 1, \ldots, j-1, j+1, \ldots, n$. Let $a = a_2 + \cdots + a_n$. Then $|a|_{\alpha_1} \leqslant \max(|a_2|_{\alpha_1}, \ldots, |a_n|_{\alpha_1}) < 1$ while for $j = 2, \ldots, n$, $|a|_{\alpha_j} = \max(|a_2|_{\alpha_j}, \ldots, |a_n|_{\alpha_j}) = 1$. ‖

PROPOSITION 5. *Let $A \in F(k, \mathfrak{B})$ and let $\alpha_1, \ldots, \alpha_n$ be distinct elements of I. Then there is an element $a \in A$ such that $\mathrm{ord}_{\alpha_i} A = \mathrm{ord}_{\alpha_i} a$ for $i = 1, \ldots, n$.*

Proof. Let $1 \leqslant i \leqslant n$ and let $a_i \in A$ be such that $\mathrm{ord}_{\alpha_i} A = \mathrm{ord}_{\alpha_i} a_i$. By Proposition 4 there is an element $b_i \in \mathfrak{o}$ such that $|b_i|_{\alpha_i} = 1$ and $|b_i|_{\alpha_j} < 1$ for $j = 1, \ldots, i-1, i+1, \ldots, n$, that is $\mathrm{ord}_{\alpha_i} b_i = 0$ and $\mathrm{ord}_{\alpha_j} b_i > 0$ for $j \neq i$. Let $a = a_1 b_1 + \cdots + a_n b_n$. For each i we have $\mathrm{ord}_{\alpha_i} a_i b_i = \mathrm{ord}_{\alpha_i} a_i + \mathrm{ord}_{\alpha_i} b_i = \mathrm{ord}_{\alpha_i} a_i = \mathrm{ord}_{\alpha_i} A$, while for $j \neq i$, we have $\mathrm{ord}_{\alpha_i} a_j b_j = \mathrm{ord}_{\alpha_i} a_j + \mathrm{ord}_{\alpha_i} b_j > \mathrm{ord}_{\alpha_i} a_j \geqslant \mathrm{ord}_{\alpha_i} A$. Therefore,
$$\mathrm{ord}_{\alpha_i} a = \min(\mathrm{ord}_{\alpha_i} a_1 b_1, \ldots, \mathrm{ord}_{\alpha_i} a_n b_n) = \mathrm{ord}_{\alpha_i} A. ‖$$

PROPOSITION 6. *If $A \in F(k, \mathfrak{B})$ then $\mathrm{ord}_{\alpha} A^{-1} A = 0$ for all $\alpha \in I$.*

Proof. If $\mathrm{ord}_{\alpha} A = 0$ for all $\alpha \in I$ then $A = \mathfrak{o}$ by (D), $A^{-1} = \mathfrak{o}$, and so $A^{-1}A = \mathfrak{o}$ and the result is true in this case. Assume then that $\mathrm{ord}_{\alpha} A \neq 0$ for some $\alpha \in I$ and let $|\ |_{\alpha_1}, \ldots, |\ |_{\alpha_n}$ be those valuations in $\mathfrak{B}$ such that $\mathrm{ord}_{\alpha_i} A \neq 0$ for $i = 1, \ldots, n$. We consider two cases.

Case 1. These are all of the valuations in $\mathfrak{B}$. Let $a \in A$ be such that $\mathrm{ord}_{\alpha_i} A = \mathrm{ord}_{\alpha_i} a$ for $i = 1, \ldots, n$. We certainly have $a\mathfrak{o} \subseteq A$. Let $b \in A$: then $\mathrm{ord}_{\alpha_i} ba^{-1} = \mathrm{ord}_{\alpha_i} b - \mathrm{ord}_{\alpha_i} a \geqslant 0$ for $i = 1, \ldots, n$ and so $ba^{-1} \in \mathfrak{o}$. Then $b \in a\mathfrak{o}$. Hence we have $A = a\mathfrak{o}$. If $c \in a^{-1}\mathfrak{o}$ then $ac \in \mathfrak{o}$ and so $cA = ca\mathfrak{o} \subseteq \mathfrak{o}$: thus $a^{-1}\mathfrak{o} \subseteq A^{-1}$. If $c \in A^{-1}$ then $ca\mathfrak{o} = cA \subseteq \mathfrak{o}$ and so $ca \in \mathfrak{o}$: thus $A^{-1} \subseteq a^{-1}\mathfrak{o}$. Therefore $A^{-1} = a^{-1}\mathfrak{o}$ and we have $A^{-1}A = \mathfrak{o}$.

Case 2. $\mathfrak{B}$ has other valuations besides these. Let $|\ |_{\alpha}$ be one of these other valuations. Let $a \in A$ be such that $\mathrm{ord}_{\alpha_i} A = \mathrm{ord}_{\alpha_i} a$ for $i = 1, \ldots, n$ and $\mathrm{ord}_{\alpha} a = 0$. If there is a valuation $|\ |_{\beta} \in \mathfrak{B}$ which is different than $|\ |_{\alpha}$, $|\ |_{\alpha_1}, \ldots, |\ |_{\alpha_n}$ and such that $|a|_{\beta} \neq 1$, then there are only a finite number of these by (B), and we let these be $|\ |_{\alpha_i}$ for $i = n+1, \ldots, r$. Let $n+1 \leqslant i \leqslant r$. By Proposition 4 there is an element $a_i \in \mathfrak{o}$ such that $|a_i|_{\alpha_i} < 1$ and $|a_i|_{\alpha_j} = 1$ for $j = 1, \ldots, n, \ldots, i-1, i+1, \ldots, r$. Let $b = (a_{n+1} \cdots a_r)^m$. For large enough m we will have $|b|_{\alpha_i} \leqslant |a|_{\alpha_i}$. If $|\ |_{\gamma}$ is any valuation in $\mathfrak{B}$ other than $|\ |_{\alpha}, |\ |_{\alpha_1}, \ldots, |\ |_{\alpha_r}$ then $|a|_{\gamma} = 1$ and $|b|_{\gamma} \leqslant 1$ since $b \in \mathfrak{o}$.

If there is no valuation $|\ |_{\beta} \in \mathfrak{B}$ that is different than $|\ |_{\alpha}, |\ |_{\alpha_1}, \ldots, |\ |_{\alpha_n}$ and such that $|a|_{\beta} \neq 1$, we take $b = 1$. In both situations we have
$$\mathrm{ord}_{\alpha} a^{-1} b = 0,$$
$$\mathrm{ord}_{\alpha_i} a^{-1} b = \mathrm{ord}_{\alpha_i} a^{-1} = -\mathrm{ord}_{\alpha_i} a = -\mathrm{ord}_{\alpha_i} A \quad \text{for } i = 1, \ldots, n,$$
$$\mathrm{ord}_{\beta} a^{-1} b \geqslant 0 \quad \text{for all valuations } |\ |_{\beta} \text{ in } \mathfrak{B} \text{ other than } |\ |_{\alpha}, |\ |_{\alpha_1}, \ldots, |\ |_{\alpha_n}.$$

If $a' \in A$ then for all $\gamma \in I$,

$$\operatorname{ord}_\gamma a'a^{-1}b = \operatorname{ord}_\gamma a' + \operatorname{ord}_\gamma a^{-1}b$$
$$\geqslant \operatorname{ord}_\gamma A - \operatorname{ord}_\gamma A = 0,$$

so that $a'a^{-1}b \in \mathfrak{o}$. This means that $a^{-1}b \in A^{-1}$ and so $b = (a^{-1}b)a \in A^{-1}A$. We have $|b|_\alpha = 1$ and $|b|_{\alpha_i} = 1$ for $i = 1, \ldots, n$, that is, $\operatorname{ord}_\alpha b = 0$ and $\operatorname{ord}_{\alpha_i} b = 0$ for $i = 1, \ldots, n$. Hence $\operatorname{ord}_\alpha A^{-1}A \leqslant 0$ and $\operatorname{ord}_{\alpha_i} A^{-1}A \leqslant 0$ for $i = 1, \ldots, n$. But $A^{-1}A \subseteq \mathfrak{o}$ and so we conclude that $\operatorname{ord}_\alpha A^{-1}A = 0$ and $\operatorname{ord}_{\alpha_i} A^{-1}A = 0$ for $i = 1, \ldots, n$. Now, $|\ |_\alpha$ was an arbitrary valuation in $\mathfrak{B}$ other than $|\ |_{\alpha_1}, \ldots, |\ |_{\alpha_n}$ and so we have $\operatorname{ord}_\alpha A^{-1}A = 0$ for all $\alpha \in I$. $\|$

COROLLARY. *If $A \in F(k,\mathfrak{B})$ and $\alpha \in I$ then $\operatorname{ord}_\alpha A^{-1} = -\operatorname{ord}_\alpha A$.*

THEOREM 1. *If $(k,\mathfrak{B})$ is a Dedekind field then $F(k,\mathfrak{B})$ is a group.*

Let $(k,\mathfrak{B})$ be a Dedekind field. We shall now give a more complete description of the group $F(k,\mathfrak{B})$. An Abelian group G (written multiplicatively) is called a *free Abelian group* if there is a subset S of G such that every element of G can be written uniquely in the form $\prod_{s \in S} s^{n_s}$, where n_s is an integer for each $s \in S$ and $n_s \neq 0$ for only a finite number of $s \in S$, aside from the order in which the factors are written. The set S is called a *system of free generators* of G. We shall show that $F(k,\mathfrak{B})$ is a free Abelian group with the set of non-zero prime ideals of $\mathfrak{o}$ as a system of free generators. In proving this we shall use the following

LEMMA. *For each $\alpha \in I$, $\operatorname{ord}_\alpha \mathfrak{p}_\alpha = 1$.*

Proof. Our claim is that there is a prime element of k with respect to the valuation $|\ |_\alpha$ which is in $\mathfrak{o}$. Let π be a prime element of k with respect to $|\ |_\alpha$. If $\pi \in \mathfrak{p}_\alpha$ we are finished: however, we may have $\pi \notin \mathfrak{o}$. Let $|\ |_{\alpha_1}, \ldots,$ $|\ |_{\alpha_n}$ be those valuations in $\mathfrak{B}$, other than $|\ |_\alpha$, such that $|\pi|_{\alpha_i} \neq 1$ for $i = 1, \ldots,$ n. By Proposition 4 there is an element $a \in \mathfrak{o}$ such that $|a|_\alpha = 1$ and $|a|_{\alpha_i} < 1$ for $i = 1, \ldots, n$. Then, for sufficiently large positive integers s, we have $|a^s\pi|_\alpha = |\pi|_\alpha$, $|a^s\pi|_{\alpha_i} = |a|_{\alpha_i}{}^s |\pi|_{\alpha_i} < 1$ for $i = 1, \ldots, n$, and $|a^s\pi|_\beta \leqslant 1$ for $\beta \neq \alpha, \alpha_1, \ldots, \alpha_n$. Hence $a^s\pi \in \mathfrak{o}$ and this element is a prime element with respect to $|\ |_\alpha$. $\|$

THEOREM 2. *If $(k,\mathfrak{B})$ is a Dedekind field and if $A \in F(k,\mathfrak{B})$ then*

$$A = \prod_{\alpha \in I} \mathfrak{p}_\alpha^{\operatorname{ord}_\alpha A},$$

and this is the only way of writing A as a product of powers of the nonzero prime ideals of $\mathfrak{o}$, aside from the order in which the factors are written.

Proof. Let

$$A \prod_{\alpha \in I} \mathfrak{p}_\alpha^{-\operatorname{ord}_\alpha A} = B.$$

By Proposition 2 and the corollary to Proposition 6 we have

$$\text{ord}_\beta \, B = \text{ord}_\beta \, A - \sum_{\alpha \in I} \text{ord}_\alpha \, A \, \text{ord}_\beta \, \mathfrak{p}_\alpha$$

for all $\beta \in I$. If $\alpha \neq \beta$ it follows from (C) that $\text{ord}_\beta \, \mathfrak{p}_\alpha = 0$, while $\text{ord}_\beta \, \mathfrak{p}_\beta = 1$ by the lemma. Hence $\text{ord}_\beta \, B = 0$ for all $\beta \in I$. Therefore, $B = \mathfrak{o}$ so that $A = \Pi \mathfrak{p}_\alpha^{\text{ord}_\alpha A}$.

Suppose that

$$\prod_{\alpha \in I} \mathfrak{p}_\alpha^{r_\alpha} = \prod_{\alpha \in I} \mathfrak{p}_\alpha^{s_\alpha}$$

where the r_α and s_α are integers, only a finite number of which are not zero. Then

$$\prod_{\alpha \in I} \mathfrak{p}_\alpha^{r_\alpha - s_\alpha} = \mathfrak{o}$$

so that for all $\beta \in I$,

$$0 = \text{ord}_\beta \prod_{\alpha \in I} \mathfrak{p}_\alpha^{r_\alpha - s_\alpha},$$

$$= \sum_{\alpha \in I} (r_\alpha - s_\alpha) \, \text{ord}_\beta \, \mathfrak{p}_\alpha = r_\beta - s_\beta. \, \|$$

COROLLARY. *Let* $(k, \mathfrak{B})$ *be a Dedekind field with a ring of integers* $\mathfrak{o}$. *If* $\mathfrak{p}$ *is a nonzero prime ideal of* $\mathfrak{o}$ *then* $\mathfrak{p} = \mathfrak{p}_\alpha$ *for some* $\alpha \in I$.

Proof. By Theorem 2 we can write $\mathfrak{p}$ as a product of certain of the $\mathfrak{p}_\alpha$. Then by the defining property of prime ideals one of these factors, say $\mathfrak{p}_\alpha$, is contained in $\mathfrak{p}$. However, $\mathfrak{p}$ is the product of $\mathfrak{p}_\alpha$ and other ideals of $\mathfrak{o}$ so that $\mathfrak{p} \subseteq \mathfrak{p}_\alpha$. Therefore $\mathfrak{p} = \mathfrak{p}_\alpha$. $\|$

It follows from (C) that there are no proper inclusion relations between the $\mathfrak{p}_\alpha$, and, as a consequence of this corollary, every nonzero prime ideal of $\mathfrak{o}$ is a maximal ideal.

If $(k, \mathfrak{B})$ is a Dedekind field and if $A, B \in F(k, \mathfrak{B})$ we can form two other fractional ideals from A and B, namely, $A + B = \{a + b \mid a \in A, b \in B\}$ and $A \cap B$. Both of these are elements of $F(k, \mathfrak{B})$, and we have the

PROPOSITION 7. *If* $A, B \in F(k, \mathfrak{B})$ *then for each* $\alpha \in I$,

$$\text{ord}_\alpha \, (A + B) = \min \, (\text{ord}_\alpha \, A, \text{ord}_\alpha \, B),$$

$$\text{ord}_\alpha \, (A \cap B) = \max \, (\text{ord}_\alpha \, A, \text{ord}_\alpha \, B).$$

Proof. We have

$$\text{ord}_\alpha \, (A + B) = \min_{c \in A + B} \text{ord}_\alpha \, c,$$

$$= \min_{a \in A, b \in B} \text{ord}_\alpha \, (a + b),$$

$$\geqslant \min_{a \in A, b \in B} \min \, (\text{ord}_\alpha \, a, \text{ord}_\alpha \, b),$$

$$= \min \, (\min_{a \in A} \text{ord}_\alpha \, a, \min_{b \in B} \text{ord}_\alpha \, b),$$

$$= \min \, (\text{ord}_\alpha \, A, \text{ord}_\alpha \, B).$$

On the other hand, let $a \in A$ be such that $\text{ord}_\alpha A = \text{ord}_\alpha a$. Since $a \in A + B$ we have $\text{ord}_\alpha (A + B) \leqslant \text{ord}_\alpha a = \text{ord}_\alpha A$. Similarly, $\text{ord}_\alpha (A + B) \leqslant \text{ord}_\alpha B$. Thus $\text{ord}_\alpha (A + B) \leqslant \min (\text{ord}_\alpha A, \text{ord}_\alpha B)$. We leave the proof of the other part of the proposition as an exercise (Exercise 5). ∥

PROPOSITION 8. *Every ideal of* $\mathfrak{o}$ *can be generated by two of its elements, that is, if* $\mathfrak{a}$ *is an ideal of* $\mathfrak{o}$ *then there are elements* $a, b \in \mathfrak{a}$ *such that* $\mathfrak{a} = a\mathfrak{o} + b\mathfrak{o}$.

Proof. This is certainly true of the zero ideal of $\mathfrak{o}$. Let $\mathfrak{a}$ be a nonzero ideal of $\mathfrak{o}$ and let $a \in \mathfrak{a}$, $a \neq 0$. If $\text{ord}_\alpha \mathfrak{a} = \text{ord}_\alpha a$ for all $\alpha \in I$, and if $b \in \mathfrak{a}$, then $\text{ord}_\alpha ba^{-1} = \text{ord}_\alpha b - \text{ord}_\alpha a \geqslant 0$ for all $\alpha \in I$ and so $b \in a\mathfrak{o}$. Thus $\mathfrak{a} \subseteq a\mathfrak{o}$. But $a\mathfrak{o} \subseteq \mathfrak{a}$ so that $\mathfrak{a} = a\mathfrak{o}$ in this case. On the other hand, there are at most a finite number of valuations in $\mathfrak{B}$, say $\mid \ \mid_{\alpha_1}, \ldots, \mid \ \mid_{\alpha_n}$, such that $\text{ord}_{\alpha_i} \mathfrak{a} \neq \text{ord}_{\alpha_i} a$ for $i = 1, \ldots, n$. By Proposition 5 there is an element $b \in \mathfrak{a}$ such that $\text{ord}_{\alpha_i} \mathfrak{a} = \text{ord}_{\alpha_i} b$ for $i = 1, \ldots, n$. Then, for each $\alpha \in I$, $\text{ord}_\alpha \mathfrak{a} = \min (\text{ord}_\alpha a, \text{ord}_\alpha b) = \min (\text{ord}_\alpha a\mathfrak{o}, \text{ord}_\alpha b\mathfrak{o}) = \text{ord}_\alpha (a\mathfrak{o} + b\mathfrak{o})$. Now it follows from Theorem 2 that an element $A \in F(k, \mathfrak{B})$ is completely determined by the set of integers $\{\text{ord}_\alpha A \mid \alpha \in I\}$. Hence $\mathfrak{a} = a\mathfrak{o} + b\mathfrak{o}$. ∥

2. Extensions of Dedekind fields

Let $(k, \mathfrak{B})$ be a Dedekind field and let K be a finite separable extension of k. By Theorem 17 of Chapter 3 each valuation in $\mathfrak{B}$ has a finite number of prolongations to valuations on K. Let $\mathfrak{W}$ be the family of valuations on K consisting of all of the prolongations of all of the valuation in $\mathfrak{B}$ to valuations on K. Then

(a) no two valuations in $\mathfrak{W}$ are equivalent, and

(b) if $\mid \ \mid \in \mathfrak{B}$ then every prolongation of $\mid \ \mid$ to a valuation on K is in $\mathfrak{W}$, and if $\mid \ \mid_1 \in \mathfrak{W}$ then the restriction of $\mid \ \mid_1$ to k is in $\mathfrak{B}$. The principal result of this section is

THEOREM 3. *The composite object* $(K, \mathfrak{W})$ *is a Dedekind field.*

Let $\mathfrak{W}$ be indexed by a set J (there is no danger in this section of confusing this set with the ring of rational integers). For each $\alpha \in J$ let $\mathfrak{O}_\alpha$ be the ring of integers of K with respect to the valuation $\mid \ \mid_\alpha$ and let

$$\mathfrak{O} = \bigcap_{\alpha \in J} \mathfrak{O}_\alpha.$$

THEOREM 4. *The ring* $\mathfrak{O}$ *is the integral closure in* K *of the ring of integers* $\mathfrak{o}$ *of* $(k, \mathfrak{B})$, *that is, an element* $a \in K$ *is in* $\mathfrak{O}$ *if and only if* a *is a root of a monic polynomial in* $\mathfrak{o}[x]$. *In fact,* $a \in \mathfrak{O}$ *if and only if* $\text{Irr} (k, a) \in \mathfrak{o}[x]$.

Proof. Let K' be a finite normal separable extension of k which contains K and let $\mathfrak{W}'$ be the family of valuations on K' which has the same relation to $\mathfrak{W}$ as $\mathfrak{W}$ has to $\mathfrak{V}$. If $\sigma \in G(K'/k)$ and $| \ | \in \mathfrak{W}'$ we define the valuation $| \ |^\sigma$ on K' by $|a|^\sigma = |\sigma(a)|$. The restrictions of $| \ |$ and $| \ |^\sigma$ to K have the same restriction to k, and so the restriction of $| \ |^\sigma$ to k belongs to $\mathfrak{V}$. Hence the restriction of $| \ |^\sigma$ to K belongs to $\mathfrak{W}$. Therefore, if $a \in \mathfrak{O}$ we have $|\sigma(a)| \leqslant 1$ for all $| \ | \in \mathfrak{W}'$ and all $\sigma \in G(K'/k)$.

Let $a \in K$ and let $f(x) = \mathrm{Irr}\,(k,a)$. Then

$$f(x) = \prod (x - \sigma(a)),$$

where σ runs through some subset of $G(K'/k)$. If $a \in \mathfrak{O}$ then by the remarks made above, every coefficient of $f(x)$ is in $\mathfrak{o}$, that is, $f(x) \in \mathfrak{o}[x]$.

Assume that $a \in K$ but $a \notin \mathfrak{O}$. Then $|a| > 1$ for some $| \ | \in \mathfrak{W}$. Let $f(x) = x^n + b_{n-1}x^{n-1} + \cdots + b_0$. Suppose that $b_0, \ldots, b_{n-1} \in \mathfrak{o}$. Then $|b_i| \leqslant 1$ for $i = 0, \ldots, n-1$, so that for each i, $|b_i a^i| < |a^n|$. Therefore, $|f(a)| = |a^n| \neq 0$, which contradicts the fact that $f(a) = 0$. Thus, if $f(x) \in \mathfrak{o}\,[x]$ we must have $a \in \mathfrak{O}$. $\|$

Before beginning the proof of Theorem 3 we shall obtain a preliminary result known as the approximation theorem. This theorem is a very important one in the global theory of valuations. As we shall see, it is the key to one phase of our proof of Theorem 3.

LEMMA. *Let* $| \ |_1, \ldots, | \ |_n$ *be a finite number of inequivalent nontrivial valuations on a field* L.[1] *Then there is an element* $a \in L$ *such that* $|a|_1 > 1$ *and* $|a|_i < 1$ *for* $i = 2, \ldots, n$.

Proof. By induction on n. Suppose that $n = 2$. Then there are elements $b, c \in L$ such that $|b|_1 < 1$, $|b|_2 \geqslant 1$, $|c|_1 \geqslant 1$, and $|c|_2 < 1$: this is because $| \ |_1$ and $| \ |_2$ are inequivalent. Let $a = b^{-1}c$: then $|a|_1 > 1$ and $|a|_2 < 1$. Suppose now that $n > 2$ and that there is an element $b \in L$ such that $|b|_1 > 1$ and $|b|_i < 1$ for $i = 2, \ldots, n-1$. Also let $c \in L$ be such that $|c|_1 > 1$ and $|c|_n < 1$.

If $|b|_n \leqslant 1$ let $a = cb^r$ where r is a positive integer. Then $|a|_1 > 1$ and $|a|_n < 1$. If $2 \leqslant i \leqslant n-1$ then $|a|_i = |c|_i |b|_i^r$ and so if we choose r large enough we will have $|a|_i < 1$.

If $|b|_n > 1$ we let $a = cb^r/(1 + b^r)$ where r is a positive integer. If $2 \leqslant i \leqslant n-1$ we have $|b|_i < 1$ and

$$\left| \frac{b^r}{1 + b^r} \right|_i = \frac{|b|_i^r}{|1 + b^r|_i} \leqslant \frac{|b|_i^r}{1 - |b|_i^r} \to 0$$

[1] In the statements of this lemma, its corollary, and the Approximation Theorem, the valuations may be completely arbitrary. The Approximation Theorem and its consequences have become very important in the past 20 years. Although special cases were known and used for many years before, the theorem was first stated and proved in its full generality in the paper by E. Artin and G. Whaples, "Axiomatic Characterization of Fields by the Product Formula for Valuations," *Bulletin of the American Mathematical Society, 51,* 1945, pp. 469–92.

as $r \to \infty$ and $|a|_i < 1$ for large r. If $j = 1$ or n we have $|b|_j > 1$ so that

$$\left| \frac{b^r}{1 + b^r} - 1 \right|_j = \frac{1}{|1 + b^r|_j} \leqslant \frac{1}{|b|_j^r - 1} \to 0$$

as $r \to \infty$. Thus $b^r/(1 + b^r) \to 1$ as $r \to \infty$, with respect to the valuation $| \ |_j$. Therefore, $|a|_1 > 1$ and $|a|_n < 1$ for large r. ‖

COROLLARY. *If $\varepsilon > 0$ then there is an element $a \in L$ such that $|a - 1|_1 < \varepsilon$ and $|a|_i < \varepsilon$ for $i = 2, \ldots, n$.*

Proof. Let $b \in L$ be such that $|b|_1 > 1$ and $|b|_i < 1$ for $i = 2, \ldots, n$, and let $a = b^r/(1 + b^r)$ where r is a positive integer. Then, as $r \to \infty$, $|a - 1|_1 \to 0$ while $|a|_i \to 0$ for $i = 2, \ldots, n$. ‖

THEOREM 5. (The Approximation Theorem.) *Let $| \ |_1, \ldots, | \ |_n$ be a finite number of inequivalent nontrivial valuations on a field L. Let $a_1, \ldots, a_n \in L$. If $\varepsilon > 0$ there is an element $a \in L$ such that $|a - a_i|_i < \varepsilon$ for $i = 1, \ldots, n$.*

Proof. If each $a_i = 0$ take $a = 0$. Otherwise, let

$$M = \max_{1 \leqslant i, j \leqslant n} |a_i|_j.$$

If $1 \leqslant j \leqslant n$ then by the above corollary there is an element $b_j \in L$ such that $|b_j - 1|_j < \varepsilon/nM$ and $|b_j|_i < \varepsilon/nM$ for $i \neq j$. Let $a = a_1 b_1 + \cdots + a_n b_n$. Then for each i,

$$|a - a_i|_i = |a_1 b_1 + \cdots + a_i(b_i - 1) + \cdots + a_n b_n|_i$$
$$\leqslant |a_1|_i |b_1|_i + \cdots + |a_i|_i |b_i - 1|_i + \cdots + |a_n|_i |b_n|_i$$
$$\leqslant nM(\varepsilon/nM) = \varepsilon. \ ‖$$

Finally we can turn to the proof of Theorem 3. We shall show in a series of propositions that the conditions (A)–(D) of the last section hold for $(K, \mathfrak{W})$.

PROPOSITION 1. *The condition* (B) *holds for* $(K, \mathfrak{W})$.

Proof. Let $a \in \mathfrak{O}$ and let

$$\text{Irr } (k, a) = x^n + b_{n-1} x^{n-1} + \cdots + b_0.$$

Let $| \ | \in \mathfrak{W}$ and assume that $|b_i| = 1$ or 0 for $i = 0, \ldots, n - 1$. Assuming that $a \neq 0$, suppose that $|a| < 1$. Then $|b_0| > |b_i a^i|$ for $i = 1, \ldots, n (b_n = 1)$ and so $0 = |a^n + b_{n-1} a^{n-1} + \cdots + b_0| = |b_0|$: this implies that $b_0 = 0$ which cannot be true since Irr (k, a) is irreducible in $k[x]$. Thus we must have $|a| = 1$. Since $|b_i| = 1$ or 0 for $i = 0, \ldots, n - 1$ for all but a finite number of valuations in $\mathfrak{B}$ we must have $|a| = 1$ for all but a finite number of valuations in $\mathfrak{W}$. ∥

PROPOSITION 2. *The condition* (A) *holds for* $(K,\mathfrak{W})$.

Proof. We must show that every element of K can be written as a quotient of elements of $\mathfrak{O}$. Let $a \in K$: if $a \in \mathfrak{O}$ then $a = a/1$, and so we assume that $a \notin \mathfrak{O}$. Let Irr $(k,a) = x^n + b_{n-1}x^{n-1} + \cdots + b_0$. Let $|\ | \in \mathfrak{W}$ and suppose that $|b_i| = 0$ or 1 for $i = 0, \ldots, n - 1$. If $|a| > 1$ then $|a^n| > |b_i a^i|$ for $i = 0, \ldots, n - 1$, and so $0 = |a^n + b_{n-1}a^{n-1} + \cdots + b_0| = |a^n|$, which is not true since $a \neq 0$. Hence $|a| = 1$. Thus, there are only a finite number of valuations in $\mathfrak{W}$, say $|\ |_{\alpha_1}, \ldots, |\ |_{\alpha_m}$, such that $|a|_{\alpha_i} \neq 1$ for $i = 1, \ldots, m$. Each of these gives one of the valuations in $\mathfrak{B}$ when restricted to k. By the lemma preceding Theorem 2 there is a prime element π_{α_i} of k with respect to $|\ |_{\alpha_i}$ in $\mathfrak{o}$, for $i = 1, \ldots, m$. Let $b = \pi_{\alpha_1} \ldots \pi_{\alpha_m}$. Then $b \neq 0$ and $|b|_{\alpha_i} < 1$ for each i. Hence, for large positive integers r we have $|ab^r|_{\alpha_i} < 1$ for $i = 1, \ldots, m$. Thus $|ab^r|_\alpha \leqslant 1$ for all $\alpha \in J$ and so $ab^r = c \in \mathfrak{O}$ and we have $a = c/b^r$. ∥

PROPOSITION 3. *The condition* (C) *holds for* $(K,\mathfrak{W})$.

Proof. Let $|\ |_\alpha$ and $|\ |_\beta$ be distinct valuations in $\mathfrak{W}$: we shall show that there is an element $a \in \mathfrak{O}$ with $|a|_\alpha < 1$ and $|a|_\beta = 1$. If $|\ |_\alpha$ and $|\ |_\beta$ are prolongations of distinct valuations in $\mathfrak{B}$ we can use the fact that (C) holds for $(k,\mathfrak{B})$ and actually choose a from $\mathfrak{o}$. Therefore, for the rest of the proof we shall assume that $|\ |_\alpha$ and $|\ |_\beta$ are prolongations of the same valuation $|\ |$ in $\mathfrak{B}$. Let $|\ |_{\alpha_1}, \ldots, |\ |_{\alpha_n}$ be all of the valuations in $\mathfrak{W}$ which are prolongations of $|\ |$. For $i = 1, \ldots, n$, let π_i be a prime element of K with respect to $|\ |_{\alpha_i}$. By the Approximation Theorem, given $\varepsilon > 0$, there are elements $a_1, \ldots, a_n \in K$ such that for each i,

$$|a_i - \pi_i|_{\alpha_i} < \varepsilon, \qquad |a_i - 1|_{\alpha_j} < \varepsilon \qquad \text{for } j \neq i.$$

If we choose $\varepsilon < 1$ then for $j \neq i$,

$$|a_i|_{\alpha_j} = |a_i - 1 + 1|_{\alpha_j} = \max(|a_i - 1|_{\alpha_j}, 1) = 1$$

while

$$|a_i|_{\alpha_i} = |a_i - \pi_i + \pi_i|_{\alpha_i} \leqslant |\pi_i|_{\alpha_i} < 1,$$

where we have used the defining property of the prime element. In particular, $|a_1|_{\alpha_1} < 1$ and $|a_1|_{\alpha_2} = 1$. However, a_1 may not lie in $\mathfrak{O}$. Suppose that it is not in $\mathfrak{O}$ and let $|\ |_{\alpha_i}$, $i = n + 1, \ldots, r$ be those valuations in $\mathfrak{W}$ such that $|a_1|_{\alpha_i} > 1$ for $i = n + 1, \ldots, r$ (we know that there are only a finite number of such valuations in $\mathfrak{W}$ by the proof of Proposition 2). These valuations are prolongations of valuations in $\mathfrak{B}$ other than $|\ |$ and so by Proposition 4 of Section 1 there is an element $b \in \mathfrak{o}$ such that $|b|_{\alpha_1} = 1$ and $|b|_{\alpha_i} < 1$ for $i = n + 1, \ldots, r$. Then, if s is a large positive integer, we have

$$|a_1 b^s|_{\alpha_1} < 1,$$

$|a_1 b^s|_{\alpha_i} = 1$ for $i = 2, \ldots, n$, since $|b|_{\alpha_i} = |b|_{\alpha_1} = 1$,

$\quad\quad |a_1 b^s|_{\alpha_i} < 1$ for $i = n + 1, \ldots, r$,

$\quad\quad |a_1 b^s|_{\gamma} \leqslant 1$ for all other $|\ |_{\gamma} \in \mathfrak{W}$.

Hence $a_1 b^s \in \mathfrak{D}$ and has the desired properties. ‖

PROPOSITION 4. *The condition* (D) *holds for* $(K, \mathfrak{W})$.

Proof. Let $\mathfrak{A}$ be an ideal of $\mathfrak{D}$ which contains a unit with respect to each valuation in $\mathfrak{W}$: we must show that $\mathfrak{A} = \mathfrak{D}$. Assume that $\mathfrak{A} \neq \mathfrak{D}$. Then there is a maximal (and so prime) ideal of $\mathfrak{D}$ which contains $\mathfrak{A}$, and each ideal which contains $\mathfrak{A}$ contains a unit with respect to each valuation in $\mathfrak{W}$. Hence we shall assume that $\mathfrak{A}$ itself is a prime ideal of $\mathfrak{D}$.

Let $a \in \mathfrak{A}$, $a \neq 0$. If $|a| = 1$ for all $|\ |$ in $\mathfrak{W}$ then $a^{-1} \in \mathfrak{D}$ and so $\mathfrak{A} = \mathfrak{D}$, contrary to assumption. Thus $|a| < 1$ for at least one $|\ |$ in $\mathfrak{W}$. Since (B) holds for $(K, \mathfrak{W})$ there are only a finite number of valuations in $\mathfrak{W}$, say $|\ |_{\alpha_1}, \ldots, |\ |_{\alpha_n}$, such that $|a|_{\alpha_i} < 1$ for $i = 1, \ldots, n$. For each i let $\mathfrak{P}_{\alpha_i}$ be the intersection with $\mathfrak{D}$ of the ideal of nonunits of the ring of integers of K with respect to $|\ |_{\alpha_i}$, and let ord_{α_i} be the ordinal function of K with respect to $|\ |_{\alpha_i}$. Then each element of $a^{-1} \prod_{i=1}^{n} \mathfrak{P}_{\alpha_i}^{\mathrm{ord}_{\alpha_i} a}$ is in $\mathfrak{D}$. Hence,

$$\prod_{i=1}^{n} \mathfrak{P}_{\alpha_i}^{\mathrm{ord}_{\alpha_i} a} \subseteq a\mathfrak{D} \subseteq \mathfrak{A}.$$

Therefore, since $\mathfrak{A}$ is a prime ideal, $\mathfrak{P}_{\alpha_i} \subseteq \mathfrak{A}$ for some i.

Thus, we see that there is a valuation $|\ |_{\alpha} \in \mathfrak{W}$ such that $\mathfrak{P}_{\alpha} \subseteq \mathfrak{A}$. This valuation, restricted to k, determines a valuation in $\mathfrak{B}$ and an associated prime ideal $\mathfrak{p}_{\alpha}$ of $\mathfrak{o}$. Let $\mathfrak{p} = \mathfrak{A} \cap \mathfrak{o}$. Then $\mathfrak{p}_{\alpha} \subseteq \mathfrak{p}$. Since $\mathfrak{A} \neq \mathfrak{D}$, $1 \notin \mathfrak{A}$, so that $\mathfrak{p} \neq \mathfrak{o}$. Then by the corollary to Theorem 2 we have $\mathfrak{p}_{\alpha} = \mathfrak{p}$. We shall now show that $\mathfrak{P}_{\alpha} = \mathfrak{A}$: this gives us the desired contradiction since $\mathfrak{P}_{\alpha}$ does not contain a unit with respect to $|\ |_{\alpha}$. Suppose that $\mathfrak{P}_{\alpha} \neq \mathfrak{A}$ and let $a \in \mathfrak{A}$ but $a \notin \mathfrak{P}_{\alpha}$. By Theorem 4, there is a monic polynomial $f(x) \in \mathfrak{o}[x]$ such that $f(a) = 0$: thus $f(a) \equiv 0 \pmod{\mathfrak{P}_{\alpha}}$. Let $x^n + b_{n-1}x^{n-1} + \cdots + b_0$ be a monic polynomial of minimal degree in $\mathfrak{o}[x]$ such that

$$a^n + b_{n-1}a^{n-1} + \cdots + b_0 \equiv 0 \pmod{\mathfrak{P}_{\alpha}}.$$

Then $b_0 = -(a^n + b_{n-1}a^{n-1} + \cdots + b_1 a) \in \mathfrak{A} \cap \mathfrak{o} = \mathfrak{p} = \mathfrak{p}_{\alpha} \subseteq \mathfrak{P}_{\alpha}$. Hence $a(a^{n-1} + b_{n-1}a^{n-2} + \cdots + b_1) \in \mathfrak{P}_{\alpha}$ and since $\mathfrak{P}_{\alpha}$ is a prime ideal and $a \notin \mathfrak{P}_{\alpha}$ we must have $a^{n-1} + b_{n-1}a^{n-2} + \cdots + b_1 \in \mathfrak{P}_{\alpha}$. This contradicts the minimality of the degree of the polynomial chosen above. Therefore we must have $\mathfrak{P}_\alpha = \mathfrak{A}$. This completes the proof of the proposition and also completes the proof of the theorem. ‖

3. Factoring of ideals in extensions

Let $(k,\mathfrak{B})$ be a Dedekind field and let K be a finite separable extension of k. Let $\mathfrak{W}$ be the family of valuations on K which satisfy (a) and (b) stated at the beginning of Section 2. The Dedekind field $(K, \mathfrak{W})$ is called an *extension* of $(k,\mathfrak{B})$. Let $\mathfrak{o}$ be the ring of integers of $(k,\mathfrak{B})$ and $\mathfrak{O}$ the ring of integers of $(K,\mathfrak{W})$.

Because of the results of Section 1 we can discontinue our use of the index set I for $\mathfrak{B}$, and index the valuations in $\mathfrak{B}$ with the nonzero prime ideals of $\mathfrak{o}$. Similar remarks apply to $(K, \mathfrak{W})$. Thus, if $\mathfrak{p}$ is a nonzero prime ideal of $\mathfrak{o}$ then there is a valuation $\mid\ \mid_\mathfrak{p} \in \mathfrak{B}$ such that $\mathfrak{p} = \mathfrak{o} \cap \mathfrak{p}'$, where $\mathfrak{p}'$ is the ideal of nonunits of the ring of integers $\mathfrak{o}_\mathfrak{p}$ of $(k, \mid\ \mid_\mathfrak{p})$. We shall denote by $k_\mathfrak{p}$ the completion of k with respect to $\mid\ \mid_\mathfrak{p}$, by $\hat{\mathfrak{o}}_\mathfrak{p}$ the ring of integers of $k_\mathfrak{p}$, and $\hat{\mathfrak{p}}$ the ideal of nonunits of $\hat{\mathfrak{o}}_\mathfrak{p}$. By Theorem 7 of Chapter 3 and Exercise 6 we have

$$\bar{k}_\mathfrak{p} = \hat{\mathfrak{o}}_\mathfrak{p}/\hat{\mathfrak{p}} \cong \mathfrak{o}_\mathfrak{p}/\mathfrak{p}' \cong \mathfrak{o}/\mathfrak{p}.$$

Let $\mathfrak{P}$ be a nonzero prime ideal of $\mathfrak{O}$. Then $\mathfrak{P} \cap \mathfrak{o}$ is a nonzero prime ideal $\mathfrak{p}$ of $\mathfrak{o}$. In fact, $\mathfrak{p}$ is the prime ideal of $\mathfrak{o}$ which is determined by the restriction of the valuation $\mid\ \mid_\mathfrak{P}$ in $\mathfrak{W}$ to k. It follows immediately from this that if a nonzero prime ideal $\mathfrak{p}$ of $\mathfrak{o}$ is given, then there are only a finite number of prime ideals $\mathfrak{P}$ of $\mathfrak{O}$ such that $\mathfrak{p} = \mathfrak{P} \cap \mathfrak{o}$, since each $\mid\ \mid_\mathfrak{P}$ is a prolongation of $\mid\ \mid_\mathfrak{p}$ to a valuation on K. Furthermore, $\mathfrak{p} = \mathfrak{P} \cap \mathfrak{o}$ if and only if $\mathfrak{p} \subseteq \mathfrak{P}$. We note that the residue class degree $f(K_\mathfrak{P}/k_\mathfrak{p})$ and the ramification $e(K_\mathfrak{P}/k_\mathfrak{p})$ are completely determined by $\mathfrak{P}$, and we shall denote these integers by $f(\mathfrak{P})$ and $e(\mathfrak{P})$, respectively. We shall assume throughout this section and the next that for each $\mathfrak{P}$, $\bar{K}_\mathfrak{P}/\bar{k}_\mathfrak{p}$ is separable.

If $\mathfrak{p}$ is a nonzero prime ideal of $\mathfrak{o}$ we may consider the (nonzero) ideal of $\mathfrak{O}$ generated by $\mathfrak{p}$, namely,

$$\mathfrak{p}\mathfrak{O} = \{\text{finite sums } \textstyle\sum a_i b_i \mid a_i \in \mathfrak{p} \text{ and } b_i \in \mathfrak{O}\}.$$

By Theorem 2 we have

(*) $$\mathfrak{p}\mathfrak{O} = \prod_\mathfrak{P} \mathfrak{P}^{r(\mathfrak{P})},$$

where the product is over all nonzero prime ideals of $\mathfrak{O}$ and the exponents $r(\mathfrak{P})$ are non-negative integers, only a finite number of which are different from zero, and which are uniquely determined by $\mathfrak{p}$ and $\mathfrak{P}$ (and so by $\mathfrak{P}$ alone). We have $r(\mathfrak{P}) > 0$ if and only if $\mathfrak{p} \subseteq \mathfrak{P}$, that is, if and only if $\mathfrak{p} = \mathfrak{P} \cap \mathfrak{o}$. In writing $\mathfrak{p}\mathfrak{O}$ in the form (*) we may neglect those powers of prime ideals for which the exponent is zero, and so write

$$\mathfrak{p}\mathfrak{O} = \prod_{\mathfrak{p}\subseteq\mathfrak{P}} \mathfrak{P}^{r(\mathfrak{P})},$$

where the product is over all prime ideals $\mathfrak{P}$ of $\mathfrak{O}$ that contain $\mathfrak{p}$.

THEOREM 6. *If $\mathfrak{p}$ is a nonzero prime ideal of $\mathfrak{o}$ then*

$$\mathfrak{p}\mathfrak{O} = \prod_{\mathfrak{p}\subseteq\mathfrak{P}} \mathfrak{P}^{e(\mathfrak{P})}.$$

Proof. We denote the ordinal function of k that corresponds to the $| \ |_{\mathfrak{p}}$ by $\mathrm{ord}_{\mathfrak{p}}$, and $\mathrm{ord}_{\mathfrak{P}}$ will have the appropriate similar meaning. By Theorem 2 we have $r(\mathfrak{P}) = \mathrm{ord}_{\mathfrak{P}} \ \mathfrak{p}\mathfrak{O}$. It is clear that if π is a prime element of k with respect to $| \ |_{\mathfrak{p}}$ then $\mathrm{ord}_{\mathfrak{P}} \ \mathfrak{p}\mathfrak{O} = \mathrm{ord}_{\mathfrak{P}} \ \pi$, and this is equal to $e(\mathfrak{P}) \ \mathrm{ord}_{\mathfrak{p}} \ \pi = e(\mathfrak{P})$ by Exercise 4 of Chapter 4. ∥

If $\mathfrak{P}_1, \ldots, \mathfrak{P}_g$ are the prime ideals of $\mathfrak{O}$ that contain $\mathfrak{p}$, it follows from the formula preceding the statement of Theorem 18 in Chapter 3 that

$$[K:k] = \sum_{i=1}^{g} e(\mathfrak{P}_i)f(\mathfrak{P}_i).$$

This gives us some information about the way in which $\mathfrak{p}\mathfrak{O}$ decomposes in $\mathfrak{O}$. For example, if $[K:k] = 2$ then $g \leqslant 2$ and one of the following occurs:

$$\mathfrak{p}\mathfrak{O} = \mathfrak{P}_1\mathfrak{P}_2, \qquad \mathfrak{p}\mathfrak{O} = \mathfrak{P}_1, \quad \text{or} \quad \mathfrak{p}\mathfrak{O} = \mathfrak{P}_1{}^2.$$

In the first case, $e(\mathfrak{P}_1) = e(\mathfrak{P}_2) = f(\mathfrak{P}_1) = f(\mathfrak{P}_2) = 1$; in the second case, $e(\mathfrak{P}_1) = 1$ and $f(\mathfrak{P}_2) = 2$; in the third case, $e(\mathfrak{P}_1) = 2$ and $f(\mathfrak{P}_1) = 1$.

There is a problem that suggests itself immediately: if $\mathfrak{p}$ is a nonzero prime ideal of $\mathfrak{o}$, which prime ideals of $\mathfrak{O}$ occur more than once in the prime decomposition (*) of $\mathfrak{p}\mathfrak{O}$ in $\mathfrak{O}$? In other words, for which prime ideals $\mathfrak{P}$ of $\mathfrak{O}$ such that $\mathfrak{p} \subseteq \mathfrak{P}$ do we have $e(\mathfrak{P}) > 1$? The remainder of this section will be devoted to the study of this question.

DEFINITION. A nonzero prime ideal $\mathfrak{P}$ of $\mathfrak{O}$ is said to be *ramified* over k if $e(\mathfrak{P}) > 1$. If $e(\mathfrak{P}) = 1$ then $\mathfrak{P}$ is said to be *unramified* over k. A nonzero prime ideal $\mathfrak{p}$ of $\mathfrak{o}$ is said to be *ramified* in K if there is a ramified prime ideal $\mathfrak{P}$ of $\mathfrak{O}$ such that $\mathfrak{p} \subseteq \mathfrak{P}$. Otherwise, $\mathfrak{p}$ is said to be *unramified* in K. Finally, K is called an *unramified extension* of k if every nonzero prime ideal of $\mathfrak{O}$ is unramified over k.

We note that K is an unramified extension of k if and only if every nonzero prime ideal of $\mathfrak{o}$ is unramified in K. For convenience we have written k in place of $(k, \mathfrak{B})$ and K in place of $(K, \mathfrak{W})$. We shall continue to follow this practice.

Let

$$\mathfrak{O}' = \{a \in K \ | \ T_{K/k}(ab) \in \mathfrak{o} \text{ for all } b \in \mathfrak{O}\}.$$

PROPOSITION 1. *The set $\mathfrak{D}'$ is a fractional ideal of K and $\mathfrak{D} \subseteq \mathfrak{D}'$.*

Proof. By Exercise 14, $\mathfrak{D} \subseteq \mathfrak{D}'$ and so $\mathfrak{D}'$ is nonzero. Using properties of the trace we see immediately that if a, $b \in \mathfrak{D}'$ and $c \in \mathfrak{D}$, then $a - b \in \mathfrak{D}'$ and $ac \in \mathfrak{D}'$. It remains to be shown that there is an element $d \in \mathfrak{D}$, $d \neq 0$, such that $d\mathfrak{D}' \subseteq \mathfrak{D}$.

Let $[K:k] = n$ and let $a_1, \ldots, a_n$ be a basis of K/k. Since (A) holds for K, each a_i is the ratio of two elements of $\mathfrak{D}$. Let a_i' be the denominator of a_i and let $a = a_1' \cdots a_n'$. Then $aa_1, \ldots, aa_n \in \mathfrak{D}$ and they are linearly independent over k, and so form a basis of K/k. Thus we may assume, in the first place, that $a_i \in \mathfrak{D}$ for $i = 1, \ldots, n$.

Now let d be the discriminant of the basis $a_1, \ldots, a_n$ of K/k, that is, $d = \det[T_{K/k}(a_i a_j)]$. Then $d \in \mathfrak{D}$ (in fact, $d \in \mathfrak{o}$) and $d \neq 0$ by Theorem 29 of Chapter 1 since K/k is separable. Let $a \in \mathfrak{D}'$ and let $a = b_1 a_1 + \cdots + b_n a_n$ where $b_1, \ldots, b_n \in k$. For $i = 1, \ldots, n$,

$$T_{K/k}(a_i a) = \sum_{j=1}^{n} b_j T_{K/k}(a_i a_j),$$

and $T_{K/k}(a_i a) \in \mathfrak{o}$. Let $A = [T_{K/k}(a_i a_j)]$ and let B be the adjoint matrix of A, so that $A^{-1} = d^{-1}B$. We have

$$\begin{bmatrix} T_{K/k}(a_1 a) \\ \cdot \\ \cdot \\ \cdot \\ T_{K/k}(a_n a) \end{bmatrix} = A \begin{bmatrix} b_1 \\ \cdot \\ \cdot \\ \cdot \\ b_n \end{bmatrix}$$

so that

$$d \begin{bmatrix} b_1 \\ \cdot \\ \cdot \\ \cdot \\ b_n \end{bmatrix} = B \begin{bmatrix} T_{K/k}(a_1 a) \\ \cdot \\ \cdot \\ \cdot \\ T_{K/k}(a_n a) \end{bmatrix}.$$

Each entry of the matrix on the right is in $\mathfrak{o}$. Hence $db_i \in \mathfrak{o}$ for $i = 1, \ldots, n$. Therefore, $da = db_1 a_1 + \cdots + db_n a_n \in \mathfrak{D}$. ∥[2]

We now set $\mathfrak{D} = \mathfrak{D}'^{-1}$ and we call $\mathfrak{D}$ the *different* of K/k. Then $\mathfrak{D}$ is an ideal of $\mathfrak{D}$ (this follows from Exercise 7) and we wish to determine the prime decomposition of $\mathfrak{D}$ in $\mathfrak{D}$, that is, we wish to determine $\mathrm{ord}_{\mathfrak{P}} \mathfrak{D}$ for each nonzero prime ideal $\mathfrak{P}$ of $\mathfrak{D}$. With each such $\mathfrak{P}$ we can associate the local different $\mathfrak{D}_{\mathfrak{P}}$, which is the different of $K_{\mathfrak{P}}/k_{\mathfrak{p}}$. By Theorem 8 of Chapter 4 we have $\mathfrak{D}_{\mathfrak{P}} = \hat{\mathfrak{P}}^{m(\mathfrak{P})}$ for some non-negative integer $m(\mathfrak{P})$.

[2] A proof like this one can also be given for Theorem 8 of Chapter 4.

THEOREM 7. *We have* $\mathfrak{D} = \prod_{\mathfrak{P}} \mathfrak{P}^{m(\mathfrak{P})}$.

Proof. Let $\mathfrak{D}_{\mathfrak{P}}^{-1} = \hat{\mathfrak{D}}_{\mathfrak{P}}' = \hat{\mathfrak{P}}^{s(\mathfrak{P})}$, so that $s(\mathfrak{P}) = -m(\mathfrak{P})$. We shall show that

$$\mathfrak{D}' = \prod_{\mathfrak{P}} \mathfrak{P}^{s(\mathfrak{P})},$$

that is, that $s(\mathfrak{P}) = \text{ord}_{\mathfrak{P}} \mathfrak{D}'$ for each nonzero prime ideal $\mathfrak{P}$ of $\mathfrak{D}$.

Let $\mathfrak{P}$ be a nonzero prime ideal of $\mathfrak{D}$ and let $\mathfrak{p} = \mathfrak{P} \cap \mathfrak{o}$. Let $a \in \mathfrak{D}'$ and let $b \in \mathfrak{D}_{\mathfrak{P}}$, the ring of integers of $(K, | \ |_{\mathfrak{P}})$. Suppose that we can show that $T_{\mathfrak{P}}(ab) \in \hat{\mathfrak{o}}_{\mathfrak{p}}$, where we have denoted the trace mapping of $K_{\mathfrak{P}}/k_{\mathfrak{p}}$ by $T_{\mathfrak{P}}$. Let $\hat{b} \in \hat{\mathfrak{D}}_{\mathfrak{P}}$. It is easy to see that $\hat{b} = \lim_{n \to \infty} b_n$ where $\{b_n\}$ is a Cauchy sequence of elements of $\mathfrak{D}_{\mathfrak{P}}$. Then $T_{\mathfrak{P}}(a\hat{b}) = \lim_{n \to \infty} T_{\mathfrak{P}}(ab_n) \in \hat{\mathfrak{o}}_{\mathfrak{p}}$ (the reader should verify this). Hence $a \in \hat{\mathfrak{D}}_{\mathfrak{P}}'$ which implies that $\text{ord}_{\mathfrak{P}} a \geqslant s(\mathfrak{P})$. Therefore, $\text{ord}_{\mathfrak{P}} \mathfrak{D}' \geqslant s(\mathfrak{P})$.

Thus, we let $a \in \mathfrak{D}'$ and we consider an element $b \in \mathfrak{D}_{\mathfrak{P}}$. Let $\mathfrak{P}_1 = \mathfrak{P}$, $\mathfrak{P}_2, \ldots, \mathfrak{P}_g$ be those nonzero prime ideals of $\mathfrak{D}$ which contain $\mathfrak{p}$. By Exercise 3 there is an element $c \in K$ such that

$$\text{ord}_{\mathfrak{P}} (c - b) \geqslant N,$$

$$\text{ord}_{\mathfrak{P}_i} c \geqslant N \text{ for } i = 2, \ldots, g,$$

$$\text{ord}_{\mathfrak{Q}} c \geqslant 0 \quad \text{for every other nonzero prime ideal } \mathfrak{Q} \text{ of } \mathfrak{D},$$

where N is a large positive integer. Since $\text{ord}_{\mathfrak{P}} c = \text{ord}_{\mathfrak{P}} ((c - b) + b) \geqslant \min (\text{ord}_{\mathfrak{P}} (c - b), \text{ord}_{\mathfrak{P}} b) \geqslant 0$, we have $c \in \mathfrak{D}$. Therefore, $T_{K/k}(ac) \in \mathfrak{o}$ and we have, using Theorem 18 of Chapter 3,

$$T_{\mathfrak{P}}(ab) = T_{K/k}(ac) - T_{K/k}(ac) + T_{\mathfrak{P}}(ab),$$

$$= T_{K/k}(ac) - \sum_{i=2}^{g} T_{\mathfrak{P}_i}(ac) + T_{\mathfrak{P}}(a(b - c)).$$

If N is large enough we will have $T_{\mathfrak{P}_i}(ac) \in \hat{\mathfrak{o}}_{\mathfrak{p}}$ for $i = 2, \ldots, g$, and $T_{\mathfrak{P}} (a(b - c)) \in \hat{\mathfrak{o}}_{\mathfrak{p}}$. Therefore, $T_{\mathfrak{P}}(ab) \in \hat{\mathfrak{o}}_{\mathfrak{p}}$.

It remains to be shown that $\text{ord}_{\mathfrak{P}} \mathfrak{D}' \leqslant s(\mathfrak{P})$. Let $c \in K$ be such that $\text{ord}_{\mathfrak{P}} c = s(\mathfrak{P})$ (such an element c always exists since $s(\mathfrak{P})$ is an integer). By Exercise 3 there is an element $a \in K$ such that $\text{ord}_{\mathfrak{P}} (a - c) > s(\mathfrak{P})$ and $\text{ord}_{\mathfrak{Q}} a \geqslant 0$ for all nonzero prime ideals $\mathfrak{Q}$ of $\mathfrak{D}$ other than $\mathfrak{P}$. Then $\text{ord}_{\mathfrak{P}} a = \text{ord}_{\mathfrak{P}} ((a - c) + c) = \min (\text{ord}_{\mathfrak{P}} (a - c), \text{ord}_{\mathfrak{P}} c) = s(\mathfrak{P})$. If we show that $a \in \mathfrak{D}'$ we will have the desired inequality. Let $b \in \mathfrak{D}$. Again, let $\mathfrak{P}_1 = \mathfrak{P}$, $\mathfrak{P}_2, \ldots, \mathfrak{P}_g$ be those nonzero prime ideals of $\mathfrak{D}$ which contain $\mathfrak{p} = \mathfrak{P} \cap \mathfrak{o}$, and let $\mathfrak{q}$ be an arbitrary nonzero prime ideal of $\mathfrak{o}$ other than $\mathfrak{p}$. First of all,

$a \in \hat{\Omega}_{\mathfrak{P}}'$ and $a \in \mathfrak{D}_{\mathfrak{P}_i}$ for $i = 2, \ldots, g$. Hence, again by Theorem 18 of Chapter 3,

$$T_{K/k}(ab) = \sum_{i=1}^{g} T_{\mathfrak{P}_i}(ab) \in \hat{\mathfrak{o}}_{\mathfrak{p}} \cap k = \mathfrak{o}_{\mathfrak{p}}.$$

Furthermore, $a \in \mathfrak{D}_{\Omega}$ if $\mathfrak{q} \subseteq \Omega$ and so

$$T_{K/k}(ab) = \sum_{\mathfrak{q} \subseteq \Omega} T_{\Omega}(ab) \in \hat{\mathfrak{o}}_{\mathfrak{q}} \cap k = \mathfrak{o}_{\mathfrak{q}}.$$

Therefore, $T_{K/k}(ab) \in (\bigcap_{\mathfrak{q}} \mathfrak{o}_{\mathfrak{q}}) \cap \mathfrak{o}_{\mathfrak{p}} = \mathfrak{o}$ and so $a \in \mathfrak{D}'$. $\parallel$

THEOREM 8. *A nonzero prime ideal $\mathfrak{P}$ of $\mathfrak{D}$ is ramified over k if and only if* $\operatorname{ord}_{\mathfrak{P}} \mathfrak{D} \geqslant 1$. *There are only a finite number of nonzero prime ideals of $\mathfrak{D}$ which are ramified over k. Finally, K/k is unramified if and only if $\mathfrak{D} = \mathfrak{D}$.*

Proof. The second and third statements of the theorem follow from the first. Suppose that $\operatorname{ord}_{\mathfrak{P}} \mathfrak{D} = 0$, that is, $m(\mathfrak{P}) = 0$. Then $\mathfrak{D}_{\mathfrak{P}} = \hat{\mathfrak{D}}_{\mathfrak{P}}$ and so, by Theorem 10 of Chapter 4, $K_{\mathfrak{P}}$ is an unramified extension of $k_{\mathfrak{p}}$. Hence $e(\mathfrak{P}) = 1$ so that $\mathfrak{P}$ is unramified over k. Conversely, if $\mathfrak{P}$ is unramified over k, then $K_{\mathfrak{P}}/k_{\mathfrak{p}}$ is unramified: hence $\mathfrak{D}_{\mathfrak{P}} = \hat{\mathfrak{D}}_{\mathfrak{P}}$, $m(\mathfrak{P}) = 0$, and it follows that $\operatorname{ord}_{\mathfrak{P}} \mathfrak{D} = 0$. $\parallel$

Although we have now answered our question concerning ramification, the criterion of Theorem 8 is not always easy to apply since it is not always easy to determine the different of an extension of a Dedekind field. We shall now obtain a second result concerning ramification which, while not giving as complete information as Theorem 8, is generally easier to apply.

We shall make use of the following mapping from the group $F(K, \mathfrak{W})$ into the group $F(k, \mathfrak{W})$. If $\mathfrak{P}$ is a nonzero prime ideal of $\mathfrak{D}$ we set

$$N_{K/k}(\mathfrak{P}) = \mathfrak{p}^{f(\mathfrak{P})} \text{ where } \mathfrak{p} = \mathfrak{P} \cap \mathfrak{o}.$$

For an arbitrary $A \in F(K, \mathfrak{W})$ we have

$$A = \prod_{\mathfrak{P}} \mathfrak{P}^{\operatorname{ord}_{\mathfrak{P}} A},$$

and we set

$$N_{K/k}(A) = \prod_{\mathfrak{P}} N_{K/k}(\mathfrak{P})^{\operatorname{ord}_{\mathfrak{P}} A} = \prod_{\mathfrak{p}} \mathfrak{p}^{\sum_{\mathfrak{p} \subseteq \mathfrak{P}} f(\mathfrak{P}) \operatorname{ord}_{\mathfrak{P}} A}.$$

Then $N_{K/k}$ is a homomorphism from $F(K, \mathfrak{W})$ into $F(k, \mathfrak{W})$ and it is clear that if $A \subseteq B$ then $N_{K/k}(A) \subseteq N_{K/k}(B)$. Furthermore, if $\mathfrak{p}$ is a nonzero prime ideal of $\mathfrak{o}$ then

$$N_{K/k}(\mathfrak{p}\mathfrak{D}) = \mathfrak{p}^{\sum_{\mathfrak{p} \subseteq \mathfrak{P}} e(\mathfrak{P}) f(\mathfrak{P})} = \mathfrak{p}^{[K:k]}.$$

We call $N_{K/k}(A)$ the *norm* of the fractional ideal A. The connection between this mapping and the concept of norm of elements is given by

PROPOSITION 2. *If $a \in K$, $a \neq 0$, then*

$$N_{K/k}(a\mathfrak{O}) = N_{K/k}(a)\mathfrak{o}.$$

Proof. We have

$$a\mathfrak{O} = \prod_{\mathfrak{P}} \mathfrak{P}^{\mathrm{ord}_{\mathfrak{P}}a}$$

and therefore

$$N_{K/k}(a\mathfrak{O}) = \prod_{\mathfrak{p}} \mathfrak{p}^{\sum_{\mathfrak{p} \subseteq \mathfrak{P}} f(\mathfrak{P})\mathrm{ord}_{\mathfrak{P}}a}.$$

Let $\mathfrak{p} \subseteq \mathfrak{P}$. Considering a as an element $K_{\mathfrak{P}}$ we have $|a|_{\mathfrak{P}}^n = |N_{\mathfrak{P}}(a)|_{\mathfrak{p}}$ where $n = [K_{\mathfrak{P}} : k_{\mathfrak{p}}] = e(\mathfrak{P})f(\mathfrak{P})$, and $N_{\mathfrak{P}}$ is the norm mapping of $K_{\mathfrak{P}}/k_{\mathfrak{p}}$. Hence $n \, \mathrm{ord}_{\mathfrak{P}} \, a = \mathrm{ord}_{\mathfrak{p}} N_{\mathfrak{P}}(a) = e(\mathfrak{P}) \, \mathrm{ord}_{\mathfrak{p}} N_{\mathfrak{P}}(a)$. Therefore, $f(\mathfrak{P}) \, \mathrm{ord}_{\mathfrak{P}} \, a = \mathrm{ord}_{\mathfrak{p}} N_{\mathfrak{P}}(a)$, and we have

$$\sum_{\mathfrak{p} \subseteq \mathfrak{P}} f(\mathfrak{P}) \, \mathrm{ord}_{\mathfrak{P}} \, a = \sum_{\mathfrak{p} \subseteq \mathfrak{P}} \mathrm{ord}_{\mathfrak{p}} \, N_{\mathfrak{P}}(a)$$

$$= \mathrm{ord}_{\mathfrak{p}} \prod_{\mathfrak{p} \subseteq \mathfrak{P}} N_{\mathfrak{P}}(a) = \mathrm{ord}_{\mathfrak{p}} N_{K/k}(a),$$

by Theorem 18 of Chapter 3. Thus the equality of the proposition holds. ‖

If $\mathfrak{D}$ is the different of K/k we set $\mathfrak{d} = N_{K/k}(\mathfrak{D})$ and we call $\mathfrak{d}$ the *discriminant of K/k*. We see immediately that

$$\mathfrak{d} = \prod_{\mathfrak{p}} \mathfrak{p}^{\sum_{\mathfrak{p} \subseteq \mathfrak{P}} f(\mathfrak{P})m(\mathfrak{P})}.$$

THEOREM 9. *A nonzero prime ideal $\mathfrak{p}$ of $\mathfrak{o}$ is ramified in K if and only if $\mathrm{ord}_{\mathfrak{p}} \, \mathfrak{d} \geqslant 1$. There are only a finite number of nonzero prime ideals of $\mathfrak{o}$ that are ramified in K. Finally, K/k is unramified if and only if $\mathfrak{d} = \mathfrak{o}$.*

Proof. Again, the second and third statements of the theorem follow from the first. Let $\mathfrak{p}$ be a nonzero prime ideal of $\mathfrak{o}$. If $\mathfrak{p}$ is unramified in K then $m(\mathfrak{P}) = 0$ for all prime ideals $\mathfrak{P}$ of $\mathfrak{D}$ such that $\mathfrak{p} \subseteq \mathfrak{P}$: hence

$$\sum_{\mathfrak{p} \subseteq \mathfrak{P}} f(\mathfrak{P})m(\mathfrak{P}) = 0$$

and so $\mathrm{ord}_{\mathfrak{p}} \, \mathfrak{d} = 0$. Conversely, if $\mathrm{ord}_{\mathfrak{p}} \, \mathfrak{d} = 0$ then for each $\mathfrak{P}$ with $\mathfrak{p} \subseteq \mathfrak{P}$ we have $f(\mathfrak{P})m(\mathfrak{P}) = 0$. Since $f(\mathfrak{P}) \geqslant 1$ we must have $m(\mathfrak{P}) = 0$. Therefore, $\mathfrak{p}$ is unramified in K. ‖

We shall now obtain a result that tells us, in many cases, how to determine the discriminant of K/k. Also, this result will justify our calling $\mathfrak{d}$ the discriminant of K/k. More precisely, we shall show the connection between $\mathfrak{d}$ and the concept of discriminant of a basis of K/k as discussed in Section 9 of Chapter 1. In Section 3 of Chapter 4 we defined a minimal basis of a finite extension of a local field. We shall use the same definition for the case at hand: a basis $a_1, \ldots, a_n$ of K/k is called a *minimal basis* if $c_1a_1 + \cdots + c_na_n \in \mathfrak{D}$, where $c_1, \ldots, c_n \in k$, if and only if $c_1, \ldots, c_n \in \mathfrak{o}$. In particular, for $i = 1, \ldots, n$, $a_i = 1a_i \in \mathfrak{D}$. We denote the discriminant of the basis $a_1, \ldots, a_n$ by $d(a_1, \ldots, a_n)$.

Let $a_1, \ldots, a_n$ and $b_1, \ldots, b_n$ be minimal bases of K/k. Then there is a nonsingular $n \times n$ matrix A over k such that

$$\begin{bmatrix} b_1 \\ \cdot \\ \cdot \\ \cdot \\ b_n \end{bmatrix} = A \begin{bmatrix} a_1 \\ \cdot \\ \cdot \\ \cdot \\ a_n \end{bmatrix}.$$

Since $b_1, \ldots, b_n \in \mathfrak{D}$, each entry of A is in $\mathfrak{o}$. Furthermore,

$$\begin{bmatrix} a_1 \\ \cdot \\ \cdot \\ \cdot \\ a_n \end{bmatrix} = A^{-1} \begin{bmatrix} b_1 \\ \cdot \\ \cdot \\ \cdot \\ b_n \end{bmatrix},$$

therefore each entry of A^{-1} is in $\mathfrak{o}$. As a result, $\det A$ is a unit in $\mathfrak{o}$. Since $d(b_1, \ldots, b_n) = (\det A)^2 d(a_1, \ldots, a_n)$, we have

$$|d(b_1, \ldots, b_n)|_{\mathfrak{p}} = |d(a_1, \ldots, a_n)|_{\mathfrak{p}}$$

for all nonzero prime ideals $\mathfrak{p}$ of $\mathfrak{o}$. Therefore,

$$\mathrm{ord}_{\mathfrak{p}} \, d(b_1, \ldots, b_n) = \mathrm{ord}_{\mathfrak{p}} \, d(a_1, \ldots, a_n).$$

We note that since K/k is separable, $d(a_1, \ldots, a_n) \neq 0$.

The definition and discussion of complementary basis given in Section 3 of Chapter 4 apply equally well in the present situation. In particular, if $a_1, \ldots, a_n$ is a minimal basis of K/k, and if $a_1', \ldots, a_n'$ is its complementary basis, then $\mathfrak{D}' = a_1'\mathfrak{o} + \cdots + a_n'\mathfrak{o}$. Let

$$\begin{bmatrix} a_1' \\ \cdot \\ \cdot \\ \cdot \\ a_n' \end{bmatrix} = C \begin{bmatrix} a_1 \\ \cdot \\ \cdot \\ \cdot \\ a_n \end{bmatrix}, \quad C = [c_{ij}],$$

so that for $i = 1, \ldots, n$,

$$a_i' = \sum_{h=1}^{n} c_{ih}a_h.$$

Then for each i and j,

$$T_{K/k}(a_i' a_j') = \sum_{h=1}^{n} c_{ih} T_{K/k}(a_h a_j') = c_{ij}.$$

Hence, $d(a_1', \ldots, a_n') = d(a_1', \ldots, a_n')^2 d(a_1, \ldots, a_n)$, or $d(a_1', \ldots, a_n')$
$d(a_1, \ldots, a_n) = 1$.

THEOREM 10. *If $a_1, \ldots, a_n$ is a minimal basis of K/k then $\mathfrak{d} = d(a_1, \ldots, a_n)\mathfrak{o}$.*

Proof. Let $a_1', \ldots, a_n'$ be the complementary basis of $a_1, \ldots, a_n$. We shall show that $N_{K/k}(\mathfrak{O}') = d(a_1', \ldots, a_n')\mathfrak{o}$. From this it follows that $\mathfrak{d} = N_{K/k}(\mathfrak{O}'^{-1}) = d(a_1', \ldots, a_n')^{-1}\mathfrak{o} = d(a_1, \ldots, a_n)\mathfrak{o}$.

Let $\mathfrak{P}_1, \ldots, \mathfrak{P}_t$ be all of the nonzero prime ideals of $\mathfrak{O}$ which contain nonzero prime ideals $\mathfrak{p}$ of $\mathfrak{o}$ such that either $\mathrm{ord}_\mathfrak{p} d(a_1', \ldots, a_n') \neq 0$ or $\mathrm{ord}_\mathfrak{p} N_{K/k}(\mathfrak{O}') \neq 0$. Then $\mathrm{ord}_\mathfrak{Q} \mathfrak{O}' = 0$ for every other nonzero prime ideal $\mathfrak{Q}$ of $\mathfrak{O}$. By Proposition 5 of Section 1 there are elements $a, b \in K$ with $b \neq 0$ such that

$$\mathrm{ord}_{\mathfrak{P}_i} a = \mathrm{ord}_{\mathfrak{P}_i} \mathfrak{O}', \qquad i = 1, \ldots, t,$$
$$\mathrm{ord}_\mathfrak{Q} a \geqslant \mathrm{ord}_\mathfrak{Q} \mathfrak{O}', \qquad \text{for } \mathfrak{Q} \neq \mathfrak{P}_i,$$
$$\mathrm{ord}_{\mathfrak{P}_i} b^{-1} = \mathrm{ord}_{\mathfrak{P}_i} \mathfrak{O}'^{-1}, \qquad i = 1, \ldots, t,$$
$$\mathrm{ord}_\mathfrak{Q} b^{-1} \geqslant \mathrm{ord}_\mathfrak{Q} \mathfrak{O}'^{-1}, \qquad \text{for } \mathfrak{Q} \neq \mathfrak{P}_i.$$

Hence

$$\mathrm{ord}_{\mathfrak{P}_i} b = \mathrm{ord}_{\mathfrak{P}_i} \mathfrak{O}', \qquad i = 1, \ldots, t,$$
$$\mathrm{ord}_\mathfrak{Q} b \leqslant \mathrm{ord}_\mathfrak{Q} \mathfrak{O}', \qquad \text{for } \mathfrak{Q} \neq \mathfrak{P}_i.$$

Then $a \mathfrak{O} \subseteq \mathfrak{O}' \subseteq b \mathfrak{O}$ and so $N_{K/k}(a\mathfrak{O}) \subseteq N_{K/k}(\mathfrak{O}') \subseteq N_{K/k}(b\mathfrak{O})$. Hence, by Proposition 2, $N_{K/k}(a)\mathfrak{o} \subseteq N_{K/k}(\mathfrak{O}') \subseteq N_{K/k}(b)\mathfrak{o}$. It follows that for each nonzero prime ideal $\mathfrak{p}$ of $\mathfrak{o}$ we have $\mathrm{ord}_\mathfrak{p} N_{K/k}(b) \leqslant \mathrm{ord}_\mathfrak{p} N_{K/k} \mathfrak{O}' \leqslant \mathrm{ord}_\mathfrak{p} N_{K/k}(a)$.

Now, $aa_1, \ldots, aa_n$ is a basis of K/k and

$$d(aa_1, \ldots, aa_n) = N_{K/k}(a)^2 d(a_1, \ldots, a_n).$$

Since $a \mathfrak{O} \subseteq \mathfrak{O}'$, each aa_i can be written as a linear combination of $a_1', \ldots, a_n'$, with coefficients in $\mathfrak{o}$, so that there is an element $c_1 \in \mathfrak{o}$ such that

$$N_{K/k}(a)^2 d(a_1, \ldots, a_n) = c_1^2 d(a_1', \ldots, a_n')$$
$$= c_1^2 d(a_1', \ldots, a_n')^2 d(a_1, \ldots, a_n).$$

Hence $\mathrm{ord}_\mathfrak{p} N_{K/k}(a) \geqslant \mathrm{ord}_\mathfrak{p} d(a_1', \ldots, a_n')$. Also, $b^{-1}\mathfrak{O}' \subseteq \mathfrak{O}$, and each $b^{-1}a_i'$ can be written as a linear combination of $a_1, \ldots, a_n$, with coefficients in $\mathfrak{o}$. Thus there is an element $c_2 \in \mathfrak{o}$ such that

$$d(a_1', \ldots, a_n') = c_2^2 N_{K/k}(b)^2 d(a_1, \ldots, a_n).$$

Hence $\mathrm{ord}_\mathfrak{p} d(a_1', \ldots, a_n') \geqslant \mathrm{ord}_\mathfrak{p} N_{K/k}(b)$.

For $i = 1, \ldots, t$, $\mathrm{ord}_{\mathfrak{P}_i} a = \mathrm{ord}_{\mathfrak{P}_i} b$. Hence, for all nonzero prime ideals $\mathfrak{p}$ of $\mathfrak{o}$ such that either $\mathrm{ord}_{\mathfrak{p}} d(a_1', \ldots, a_n') \neq 0$ or $\mathrm{ord}_{\mathfrak{p}} N_{K/k}(\mathfrak{D}') \neq 0$, we have $\mathrm{ord}_{\mathfrak{p}} N_{K/k}(a) = \mathrm{ord}_{\mathfrak{p}} N_{K/k}(b)$, and $\mathrm{ord}_{\mathfrak{p}} N_{K/k}(\mathfrak{D}') = \mathrm{ord}_{\mathfrak{p}} d(a_1', \ldots, a_n')$. Both of these ordinals are equal to zero for all other $\mathfrak{p}$, and we obtain the desired result. $\|$

We note that Theorem 8 does give more complete information than Theorem 9. For example, suppose $\mathfrak{p}$ is a nonzero prime ideal of $\mathfrak{o}$ and suppose that $\mathfrak{p}\mathfrak{D} = \mathfrak{P}^{e(\mathfrak{P})}\mathfrak{Q}^{e(\mathfrak{Q})}$. Theorem 9 tells us whether or not at least one of $e(\mathfrak{P})$ and $e(\mathfrak{Q})$ is greater than one, while Theorem 8 tells us which, if either, of $e(\mathfrak{P})$ and $e(\mathfrak{Q})$ is greater than one. However, it is generally easier to determine the discriminant of K/k than to determine the different of K/k. There is one outstanding case when Theorems 8 and 9 give the same information, and we shall consider this case in the next section.

4. Galois extensions of Dedekind fields

We retain the notation of the last section, and we now assume that K is a finite Galois extension of k. Let $G = G(K/k)$. If S is a subset of K and if $\sigma \in G$ we set $\sigma(S) = \{\sigma(a) \mid a \in S\}$. If $a \in K$ and if $\sigma \in G$ then $\mathrm{Irr}\,(k,a) = \mathrm{Irr}\,(k,\sigma(a))$. Hence, by Theorem 4, $\sigma(\mathfrak{D}) = \mathfrak{D}$. If A is a fractional ideal of K, so is $\sigma(A)$, and if $A \subseteq \mathfrak{D}$ then $\sigma(A) \subseteq \mathfrak{D}$.

Let $\mathfrak{p}$ be a nonzero prime ideal of $\mathfrak{o}$ and let $\mathfrak{P}$ be a prime ideal of $\mathfrak{D}$ such that $\mathfrak{p} \subseteq \mathfrak{P}$. If $\sigma \in G$ then $\sigma(\mathfrak{P})$ is also a prime ideal of $\mathfrak{D}$ and $\mathfrak{p} \subseteq \sigma(\mathfrak{P})$.

LEMMA. *Let $\mathfrak{p}_1, \ldots, \mathfrak{p}_n$ be prime ideals of a commutative ring R and let $\mathfrak{a}$ be an ideal of R such that $\mathfrak{a} \nsubseteq \mathfrak{p}_i$ for $i = 1, \ldots, n$. Then $\mathfrak{a} \nsubseteq \bigcup_{i=1}^{n} \mathfrak{p}_i$.*

Proof. By induction on n. If $n = 1$ there is nothing to prove. Suppose that $n > 1$ and that the lemma holds for fewer than n prime ideals. Then for each i there is an $a_i \in \mathfrak{a}$ such that $a_i \notin \mathfrak{p}_j$ for $j \neq i$. If $a_i \notin \mathfrak{p}_i$ for some i then we have completed the proof. Assume that $a_i \in \mathfrak{p}_i$ for each i. Let $b_i = a_1 \cdots a_{i-1}a_{i+1} \cdots a_n$ and $b = b_1 + \cdots + b_n$. Then $b \in \mathfrak{a}$. However, if $1 \leqslant i \leqslant n$, $b_i \notin \mathfrak{p}_i$ but $b_i \in \mathfrak{p}_j$ for $j \neq i$. Therefore, $b \notin \mathfrak{p}_i$ for $i = 1, \ldots, n$. $\|$

THEOREM 11. *Let $\mathfrak{p}$ be a nonzero prime ideal of $\mathfrak{o}$ and let $\mathfrak{P}_1, \ldots, \mathfrak{P}_g$ be those prime ideals of $\mathfrak{D}$ which contain $\mathfrak{p}$. If $1 \leqslant i, j \leqslant n$ then there is an element $\sigma \in G$ such that $\sigma(\mathfrak{P}_i) = \mathfrak{P}_j$.*

Proof. We may assume that $i = 1$ and that $j \neq i$, for if $j = i$, we may take for σ the identity element of G. Suppose that $\sigma(\mathfrak{P}_1) \neq \mathfrak{P}_j$ for all $\sigma \in G$. Then $\mathfrak{P}_1 \neq \sigma(\mathfrak{P}_j)$ for all $\sigma \in G$, and since there are no proper inclusion relations between nonzero prime ideals of $\mathfrak{D}$, we have $\mathfrak{P}_1 \nsubseteq \sigma(\mathfrak{P}_j)$ for all

$\sigma \in G$. Hence, by the lemma,

$$\mathfrak{P}_1 \nsubseteq \bigcup_{\sigma \in G} \sigma(\mathfrak{P}_j).$$

Let $a \in \mathfrak{P}_1$ be such that $a \notin \sigma(\mathfrak{P}_j)$ for all $\sigma \in G$. Then $\sigma(a) \notin \mathfrak{P}_j$ for all $\sigma \in G$. However,

$$N_{K/k}(a) = \prod_{\sigma \in G} \sigma(a) \in \mathfrak{P}_1 \cap \mathfrak{o} = \mathfrak{p} \subseteq \mathfrak{P}_j,$$

which contradicts the fact that $\mathfrak{P}_j$ is a prime ideal of $\mathfrak{O}$. ‖

COROLLARY. *Let $\mathfrak{p}$ be a nonzero prime ideal of $\mathfrak{o}$ and let $\mathfrak{P}$ be a prime ideal of $\mathfrak{O}$ which contains $\mathfrak{p}$. Then $e(\mathfrak{P})$ is independent of $\mathfrak{P}$, that is, it depends only on $\mathfrak{p}$. If we denote this integer by $e(\mathfrak{p})$ we have*

$$\mathfrak{p}\mathfrak{O} = \left(\prod_{\mathfrak{p} \subseteq \mathfrak{P}} \mathfrak{P} \right)^{e(\mathfrak{p})}.$$

Furthermore, $f(\mathfrak{P})$ is independent of $\mathfrak{P}$, and we denote this integer by $f(\mathfrak{p})$. If $g(\mathfrak{p})$ is the number of prime ideals of $\mathfrak{O}$ which contain $\mathfrak{p}$, then

$$[K:k] = e(\mathfrak{p})f(\mathfrak{p})g(\mathfrak{p}).$$

Proof. Let $\mathfrak{P}_i$ and $\mathfrak{P}_j$ be prime ideals of $\mathfrak{O}$ which contain $\mathfrak{p}$. Let $\sigma \in G$ be such that $\sigma(\mathfrak{P}_i) = \mathfrak{P}_j$. Then $\mathfrak{P}_j$ appears in the prime decomposition of $\mathfrak{p}\mathfrak{O}$ exactly $e(\mathfrak{P}_j)$ times, and in the prime decomposition of $\sigma(\mathfrak{p}\mathfrak{O}) = \mathfrak{p}\mathfrak{O}$ exactly $e(\mathfrak{P}_i)$ times. Hence $e(\mathfrak{P}_i) = e(\mathfrak{P}_j)$. To show that $f(\mathfrak{P}_i) = f(\mathfrak{P}_j)$ we shall show that there is a $k_\mathfrak{p}$-isomorphism from $K_{\mathfrak{P}_i}$ onto $K_{\mathfrak{P}_j}$. If $a \in \mathfrak{O}$ we set $\sigma'(a + \mathfrak{P}_i) = \sigma(a) + \mathfrak{P}_j$. If $a + \mathfrak{P}_i = b + \mathfrak{P}_i$ then $a - b \in \mathfrak{P}_i$ and so $\sigma(a) - \sigma(b) = \sigma(a - b) \in \mathfrak{P}_j$: hence $\sigma(a) + \mathfrak{P}_j = \sigma(b) + \mathfrak{P}_j$. Thus σ' is well-defined and it is clear that σ' is the desired $k_\mathfrak{p}$-isomorphism. The last statement of the corollary follows from the formula after the proof of Theorem 6. ‖

If $\mathfrak{P}$ is a nonzero prime ideal of $\mathfrak{O}$ we set

$$Z(\mathfrak{P}) = \{\sigma \in G \mid \sigma(\mathfrak{P}) = \mathfrak{P}\},$$

and for $i \geqslant 0$,

$$W_i(\mathfrak{P}) = \{\sigma \in G \mid \sigma(a) \equiv a \,(\text{mod } \mathfrak{P}^{i+1}) \text{ for all } a \in \mathfrak{O}\}.$$

It is clear that each $W_i(\mathfrak{P})$ is a normal subgroup of $Z(\mathfrak{P})$ (see Proposition 1 of Section 5 of Chapter 4), and that

$$G \supseteq Z(\mathfrak{P}) \supseteq W_0(\mathfrak{P}) \supseteq W_1(\mathfrak{P}) \supseteq W_2(\mathfrak{P}) \supseteq \cdots.$$

The group $Z(\mathfrak{P})$ is called the *decomposition group* of $\mathfrak{P}$, while $W_0(\mathfrak{P})$ is called the *inertia group* of $\mathfrak{P}$ and the $W_i(\mathfrak{P})$, for $i \geqslant 1$, the (higher) *ramification groups* of $\mathfrak{P}$. We denote the fixed field of $Z(\mathfrak{P})$ by $D(\mathfrak{P})$, that of $W_0(\mathfrak{P})$ by

$T(\mathfrak{P})$, and that of $W_i(\mathfrak{P})$, for $i \geqslant 1$, by $V_i(\mathfrak{P})$. Each of these fields is called by the same name as the corresponding subgroup of G, that is, $D(\mathfrak{P})$ is called the *decomposition field* of $\mathfrak{P}$, and so on. We have

$$k \subseteq D(\mathfrak{P}) \subseteq T(\mathfrak{P}) \subseteq V_1(\mathfrak{P}) \subseteq V_2(\mathfrak{P}) \subseteq \cdots \subseteq K.$$

PROPOSITION 1. *If $\mathfrak{P}$ is a nonzero prime ideal of $\mathfrak{O}$, and if $\sigma \in G$, then*

$$Z(\sigma(\mathfrak{P})) = \sigma Z(\mathfrak{P})\sigma^{-1}.$$

Proof. If $\tau \in G$ then $\tau \in Z(\sigma(\mathfrak{P}))$ if and only if $\tau\sigma(\mathfrak{P}) = \sigma(\mathfrak{P})$, that is, if and only if $\sigma^{-1}\tau\sigma(\mathfrak{P}) = \mathfrak{P}$, that is, if and only if $\sigma^{-1}\tau\sigma \in Z(\mathfrak{P})$. ‖

COROLLARY. *Let $\mathfrak{P}$ be a nonzero prime ideal of $\mathfrak{O}$ and let $\mathfrak{p} = \mathfrak{P} \cap \mathfrak{o}$. If K is an Abelian extension of k then $Z(\mathfrak{P})$ depends only on $\mathfrak{p}$.*

PROPOSITION 2. *If $\mathfrak{P}$ is a nonzero prime ideal of $\mathfrak{O}$ then $Z(\mathfrak{P}) \cong G(K_{\mathfrak{P}}/k_{\mathfrak{p}})$ where $\mathfrak{p} = \mathfrak{P} \cap \mathfrak{o}$.*

Proof. If $\sigma \in G(K_{\mathfrak{P}}/k_{\mathfrak{p}})$ then the restriction of σ to K is an element σ' of G. Since $\sigma(\hat{\mathfrak{P}}) = \hat{\mathfrak{P}}$ we have $\sigma' \in Z(\mathfrak{P})$, and the mapping $\sigma \to \sigma'$ is certainly a homomorphism from $G(K_{\mathfrak{P}}/k_{\mathfrak{p}})$ into $Z(\mathfrak{P})$. Now let $\tau \in Z(\mathfrak{P})$ and let $a \in K_{\mathfrak{P}}$. Then there is a Cauchy sequence $\{a_n\}$ of elements of K such that $\lim_{n \to \infty} a_n = a$, where the convergence is with respect to $| \ |_{\mathfrak{P}}$. It follows that $\{\tau(a_n)\}$ is a Cauchy sequence of elements of K and we set $\lim_{n \to \infty} \tau(a_n) = \sigma(a)$. Suppose that $\{b_n\}$ is another Cauchy sequence of elements of K such that $\lim_{n \to \infty} b_n = a$. Then $\{a_n - b_n\}$ is a null sequence and so $\{\tau(a_n - b_n)\}$ is a null sequence (with respect to $| \ |_{\mathfrak{P}}$). Therefore $\lim_{n \to \infty} \tau(b_n) = \lim_{n \to \infty} \tau(a_n)$. Thus σ is well-defined, and $\sigma \in G(K_{\mathfrak{P}}/k_{\mathfrak{p}})$. Furthermore, $\sigma \to \sigma' = \tau$. Finally, if σ' is the identity element of $Z(\mathfrak{P})$, it follows from the fact that every element of $K_{\mathfrak{P}}$ is a Cauchy sequence of elements of K, that σ is the identity element of $G(K_{\mathfrak{P}}/k_{\mathfrak{p}})$. Thus, the mapping $\sigma \to \sigma'$ is the desired isomomorphism. ‖

From this proposition we obtain immediately

PROPOSITION 3. *For $i \geqslant 0$ we have $W_i(\mathfrak{P}) \cong W_i(K_{\mathfrak{P}}/k_{\mathfrak{p}})$.*

PROPOSITION 4. *Let $\mathfrak{P}$ be a nonzero prime ideal of $\mathfrak{O}$. Then $[D(\mathfrak{P}):k] = g(\mathfrak{p})$ and $[T(\mathfrak{P}):D(\mathfrak{P})] = f(\mathfrak{p})$, where $\mathfrak{p} = \mathfrak{P} \cap \mathfrak{o}$. Let $e(\mathfrak{p}) = dp^r$ where $p = $ char $k_{\mathfrak{p}}$ and $p \nmid d$ (p^r is replaced by 1 when char $k_{\mathfrak{p}} = 0$). Then $[V_1(\mathfrak{P}):T(\mathfrak{P})] = d$ and $[K:V_1(\mathfrak{P})] = p^r$.*

Proof. We have $[K:D(\mathfrak{P})] = \#Z(\mathfrak{P}) = \#G(K_{\mathfrak{P}}/k_{\mathfrak{p}}) = e(\mathfrak{p})f(\mathfrak{p})$, and so $[D(\mathfrak{P}):k] = g(\mathfrak{p})$ by the corollary to Theorem 11. The rest of the proposition follows from the results of Sections 2 and 5 of Chapter 4. ‖

We may now describe the successive steps in which the ideal $\mathfrak{p}\mathfrak{O}$ of $\mathfrak{O}$ decomposes into the product of prime ideals in $\mathfrak{O}$, where $\mathfrak{p}$ is a nonzero prime ideal of $\mathfrak{o}$. We focus our attention on one of the prime ideals $\mathfrak{P}$ of $\mathfrak{O}$ that contain $\mathfrak{p}$. Let $\mathfrak{O}_D$, $\mathfrak{O}_T$, and $\mathfrak{O}_V$ be the rings of integers of $D(\mathfrak{P})$, $T(\mathfrak{P})$, and $V_1(\mathfrak{P})$, respectively, so that $\mathfrak{O}_D = \mathfrak{O} \cap D(\mathfrak{P})$, etc. Let $\mathfrak{P}_D = \mathfrak{P} \cap D(\mathfrak{P})$, $\mathfrak{P}_T = \mathfrak{P} \cap T(\mathfrak{P})$, and $\mathfrak{P}_V = \mathfrak{P} \cap V_1(\mathfrak{P})$. In $\mathfrak{O}_D$, $\mathfrak{P}_D$ appears as a prime factor of $\mathfrak{p}\mathfrak{O}_D$ with exponent one. Furthermore, $[\mathfrak{O}_D/\mathfrak{P}_D : \mathfrak{o}/\mathfrak{p}] = 1$ (why?). As we move up to $T(\mathfrak{P})$ there continues to be no ramification but all enlargement of the residue class field that will take place, does take place. In other words, in $\mathfrak{O}_T$, $\mathfrak{P}_T$ appears as a prime factor of $\mathfrak{p}\mathfrak{O}_T$, still with exponent one, while $[\mathfrak{O}_T/\mathfrak{P}_T : \mathfrak{O}_D/\mathfrak{P}_D] = [\mathfrak{O}_T/\mathfrak{P}_T : \mathfrak{o}/\mathfrak{p}] = f(\mathfrak{p})$. As we go from $T(\mathfrak{P})$ to $V_1(\mathfrak{P})$, ramification begins to take place: $\mathfrak{P}_V$ appears as a prime factor of $\mathfrak{p}\mathfrak{O}_V$ with the exponent d of Proposition 5. Finally, in moving up from $V_1(\mathfrak{P})$ to K the rest of the ramification takes place, with $\mathfrak{P}$ appearing as a prime factor of $\mathfrak{p}\mathfrak{O}$ with exponent $e(\mathfrak{p})$.

EXERCISES

Sections 1 and 2

1. It follows from the corollary to Theorem 2 that each $\mathfrak{p}_\alpha$ is a maximal ideal of $\mathfrak{o}$. In turn, it follows from this that if $\alpha, \beta \in I$, $\alpha \neq \beta$, then $\mathfrak{p}_\alpha + \mathfrak{p}_\beta = \mathfrak{o}$. Show, as a direct consequence of this fact, that there is an element $a \in \mathfrak{o}$ such that $\text{ord}_\alpha (a - 1) \geqslant 1$ and $\text{ord}_\beta a \geqslant 1$.

2. Show that if $\alpha, \beta \in I$, $\alpha \neq \beta$, and if N_1 and N_2 are positive integers, then there is an element $a \in \mathfrak{o}$ such that $\text{ord}_\alpha (a - 1) \geqslant N_1$ and $\text{ord}_\beta a \geqslant N_2$.

3. Let $\alpha_1, \ldots, \alpha_n$ be distinct elements of I and let $N_1, \ldots, N_n$ be positive integers. Let $a_1, \ldots, a_n \in k$. Show that there is an element $a \in k$ such that

$$\text{ord}_{\alpha_i} (a - a_i) \geqslant N_i \text{ for } i = 1, \ldots, n,$$
$$\text{ord}_\beta a \geqslant 0 \text{ for } \beta \neq \alpha_1, \ldots, \alpha_n.$$

If $a_1, \ldots, a_n \in \mathfrak{o}$, show that we may choose $a \in \mathfrak{o}$. (Clearly the result of this exercise is a special case of the Approximation Theorem. However, as we have presented it, the result is a consequence of the fact that $(k, \mathfrak{B})$ is a Dedekind field. The proof of the Approximation Theorem does not depend on any such properties of the field in the statement of the theorem.)

4. Let $\alpha_1, \ldots, \alpha_n$ be distinct elements of I and let $r_1, \ldots, r_n$ be positive integers. Show that

$$\mathfrak{p}_{\alpha_1}{}^{r_1} \cdots \mathfrak{p}_{\alpha_n}{}^{r_n} = \mathfrak{p}_{\alpha_1}{}^{r_1} \cap \cdots \cap \mathfrak{p}_{\alpha_n}{}^{r_n}.$$

5. Use the result of Exercise 4 to prove the second half of Proposition 7 of Section 1.

6. For $\alpha \in I$ let $\mathfrak{o}_\alpha$ be the ring of integers of k with respect to the valuation $| \ |_\alpha$ and let $\mathfrak{p}_\alpha'$ be the ideal of nonunits of $\mathfrak{o}_\alpha$. Show that for all positive integers n we have $\mathfrak{o}_\alpha = \mathfrak{o} + \mathfrak{p}_\alpha'^n$ and $\mathfrak{o}/\mathfrak{p}_\alpha{}^n \cong \mathfrak{o}_\alpha/\mathfrak{p}_\alpha'^n$.

7. If A and B are fractional ideals of k show that $A \subseteq B$ if and only if

there is a fractional ideal C of $\mathfrak{o}$ such that $A = BC$, and that this is the case if and only if $\text{ord}_\alpha A \geqslant \text{ord}_\alpha B$ for all $\alpha \in I$.

8. Let $a \in k$. Show that if $f(a) = 0$ for some monic polynomial $f(x) \in \mathfrak{o}[x]$ then $a \in \mathfrak{o}$.

9. Let $k(x)$ be the field of rational functions over a field k. Let $\mathfrak{B} = \{| \ |_{p(x)} \ | \ p(x)$ is an irreducible polynomial in $k[x]\}$. Show that $(k(x),\mathfrak{B})$ is a Dedekind field and determine its ring of integers.

10. Let D be an integral domain and let a, $b \in D$. We say that b *divides* a, or that b is a *divisor* of a, if there is an element $c \in D$ such that $a = bc$. We call a and b *associates* if there is a unit $u \in D$ such that $a = bu$: this is the case if and only if a and b divide each other. An element of D is called a *prime* if it is not zero or a unit, and if it has no divisors other than its associates and units. In certain integral domains D the following condition holds: (*) Every element of D which is not zero or a unit can be written as a product of a finite number of primes. If (*) holds for D then D is called a *unique factorization domain* (UFD) if the factorization given by (*) is unique except for the order in which the factors are written and the replacement of prime factors by their associates. Show that if (*) holds for D then D is a UFD if and only if the following condition holds for D: if a, $b \in D$ and if p is a prime of D then p divides ab if and only if either p divides a or p divides b.

11. Let $\mathfrak{B}$ be the set of all p-adic valuations on Q. An extension $(K,\mathfrak{B})$ of $(Q,\mathfrak{B})$, where K is an algebraic extension of Q in the field of complex numbers, is called an *algebraic number field*. We shall assume always that $[K: Q]$ is finite, and we denote the ring of integers of K by $\mathfrak{O}$ (for convenience we write K instead of $(K,\mathfrak{B})$).

    (a) Show that u is a unit of $\mathfrak{O}$ if and only if $N_{K/Q}(u) = \pm 1$.

    (b) Show that if $a \in \mathfrak{O}$ and if $N_{K/Q}(a)$ is a rational prime then a is a prime of $\mathfrak{O}$.

    (c) Show that the condition (*) of Exercise 10 holds for $\mathfrak{O}$.

12. Let K be an algebraic number field. Show that $\mathfrak{O}$ is a UFD if and only if every ideal of $\mathfrak{O}$ is principal.

13. Let $| \ |_1, \ldots, | \ |_n$ be a finite number of arbitrary inequivalent nontrivial valuations on a field L. Show that if $r_1, \ldots, r_n$ are integers, and if
$$\prod_{i=1}^{n} |a|_i^{r_i} = 1 \text{ for all nonzero } a \in L, \text{ then } r_1 = \cdots = r_n = 0.$$

14. Let k be a Dedekind field and let K be a finite separable extension of k. Show that if $a \in \mathfrak{O}$ then $T_{K/k}(a) \in \mathfrak{o}$.

Sections 3 and 4

15. Let k be a Dedekind field and let K and L be finite separable extensions of k such that $k \subseteq K \subseteq L$. With subscripts indicating the various differents, show that $\mathfrak{D}_{L/k} = \mathfrak{D}_{L/K}\mathfrak{D}_{K/k}$.

16. Let k, K, and L be as in Exercise 15, and use subscripts to indicate the various discriminants. Show that $\mathfrak{d}_{L/k} = \mathfrak{d}_{K/k}^{[L:K]} N_{K/k}(\mathfrak{d}_{L/K})$.

17. Let k be a Dedekind field and let K be a finite separable extension of k. Let $[K:k] = n$ and suppose that $1, a, \ldots, a^{n-1}$ is a minimal basis of K/k. Let $f(x) = \text{Irr}\,(k,a)$. Show that $\mathfrak{D} = f'(a)\mathfrak{D}$.

18. Let k be a Dedekind field and let K be a finite separable extension of k. Let $a_1, \ldots, a_n$ be a minimal basis of K/k. Show that if $b_1, \ldots, b_n$ is a basis of K/k such that $b_i \in \mathfrak{D}$ for $i = 1, \ldots, n$, then for each nonzero prime ideal $\mathfrak{p}$ of $\mathfrak{o}$ we have $\text{ord}_{\mathfrak{p}}\, d(a_1, \ldots, a_n) \leqslant \text{ord}_{\mathfrak{p}}\, d(b_1, \ldots, b_n)$.

19. Let K be an algebraic number field and assume that $K = Q(a)$ where $a \in \mathfrak{D}$.

(a) Show that every element of $\mathfrak{D}$ can be written in the form

$$\frac{s_1 + s_2 a + \cdots + s_n a^{n-1}}{d(a)},$$

where $s_1, \ldots, s_n \in J$ and $d(a) = d(1, a, \ldots, a^{n-1})$.

(b) For $i = 0, 1, \ldots, n - 1$, let a_i be an element of $\mathfrak{D}$ such that when a_i is written in the above form, $s_j = 0$ for $j > i$, and s_i has the smallest possible positive value. Show that $a_1, \ldots, a_n$ is a minimal basis of K/Q.

20. Let K be an algebraic number field and let $a_1, \ldots, a_n$ be a basis of K/Q with $a_i \in \mathfrak{D}$ for $i = 1, \ldots, n$. Show that $a_1, \ldots, a_n$ is a minimal basis of K/Q if and only if abs $d(a_1, \ldots, a_n) \leqslant$ abs $d(b_1, \ldots, b_n)$ for every other basis $b_1, \ldots, b_n$ of K/Q with $b_i \in \mathfrak{D}$ for $i = 1, \ldots, n$. Here "abs" means ordinary absolute value.

21. Let K be an algebraic number field and assume that $K = Q(a)$ where $a \in \mathfrak{D}$. Let $f(x) = \text{Irr}\,(k,a)$ and $d(a) = d(1, a, \ldots, a^{n-1})$. Let Δ be the discriminant of a minimal basis of K/Q (by Theorem 10, $\mathfrak{d}$ is the ideal of J generated by Δ). Let p be a rational prime and assume that $d(a)/\Delta \not\equiv 0 \pmod p$. Let

$$f(x) \equiv f_1(x)^{e_1} \cdots f_s(x)^{e_s} \pmod p,$$

where each $f_i(x)$ is monic and is irreducible modulo p and $f_i = \deg f_i(x) > 0$. Finally, let $\mathfrak{P}_i = p\mathfrak{D} + f_i(a)\mathfrak{D}$ for $i = 1, \ldots, s$.

(a) Show that each element of $\mathfrak{D}/p\mathfrak{D}$ has exactly one representative of the form $h(a)$ where $h(x) \in J[x]$ and $\deg h(x) \leqslant n - 1$.

(b) Show that for $i = 1, \ldots, n$, $\mathfrak{P}_i$ is a prime ideal of $\mathfrak{D}$.

(c) Show that $p\mathfrak{D} = \mathfrak{P}_1^{e_1} \cdots \mathfrak{P}_s^{e_s}$.

(d) Show that for $i = 1, \ldots, n$, $f_i = f(\mathfrak{P}_i)$.

22. Let k, K, etc., be as in Section 4. Show that if $\mathfrak{P}$ is a nonzero prime ideal of $\mathfrak{D}$, and if $\sigma \in G$, then for $i \geqslant 0$ we have $W_i(\sigma(\mathfrak{P})) = \sigma W_i(\mathfrak{P})\sigma^{-1}$.

23. With the notation as in the discussion at the end of Section 4, show that $N_{D(\mathfrak{P})/k}(\mathfrak{P}_D) = \mathfrak{p}$, $N_{T(\mathfrak{P})/D(\mathfrak{P})}(\mathfrak{P}_T) = \mathfrak{P}_D^{f(\mathfrak{P})}$, and $N_{K/T(\mathfrak{P})}(\mathfrak{P}) = \mathfrak{P}_T$.

Exercises on quadratic algebraic number fields

In Exercises 24–32, K will be an algebraic number field with $[K: Q] = 2$.

24. Show that $K = Q(\sqrt{m})$ where m is a square free rational integer.

25. (a) Show that every element of K can be written uniquely in the form $(a + b\sqrt{m})/c$ where a, b, and c are rational integers, $(a,b,c) = 1$, and $c > 0$.

(b) Show that $(a + b\sqrt{m})/c \in \mathfrak{O}$ if and only if

$$c = 1 \qquad \text{when } m \not\equiv 1 \pmod{4},$$

$$c = 1, \text{ or } c = 2 \text{ and } ab \text{ odd} \qquad \text{when } m \equiv 1 \pmod 4.$$

(c) Show that

$$1, \sqrt{m} \qquad \text{when } m \not\equiv 1 \pmod 4,$$

$$1, (1 + \sqrt{m})/2 \qquad \text{when } m \equiv 1 \pmod 4,$$

form a minimal basis for K/Q.

26. Let $K = Q(\sqrt{-5})$.

(a) Show that the only units of $\mathfrak{O}$ are ± 1.

(b) Show that
$$21 = 3 \cdot 7 = (4 + \sqrt{-5})(4 - \sqrt{-5}) = (1 + 2\sqrt{-5})(1 - 2\sqrt{-5}),$$
and show that each of these six factors is a prime of $\mathfrak{O}$. Since no two of these factors are associates, this shows that $\mathfrak{O}$ is not a UFD.

(c) By Exercise 12, $\mathfrak{O}$ must have an ideal that is not principal. Show that $3\mathfrak{O} + (4 + \sqrt{-5})\mathfrak{O}$ is not a principal ideal of $\mathfrak{O}$.

27. (a) Determine the discriminant Δ of a minimal basis of K/Q.

(b) Show that if $p \mid \Delta$ then $p\mathfrak{O} = \mathfrak{P}^2$.

(c) Show that if p is odd and $p \nmid \Delta$ then $p\mathfrak{O} = \mathfrak{P}\mathfrak{Q}$ with $\mathfrak{P} \neq \mathfrak{Q}$ if m is a quadratic residue modulo p: otherwise $p\mathfrak{O} = \mathfrak{P}$.

(d) Show that if $2 \nmid \Delta$ then $2\mathfrak{O} = \mathfrak{P}\mathfrak{Q}$ with $\mathfrak{P} \neq \mathfrak{Q}$ if $m \equiv 1 \pmod 8$: otherwise $2\mathfrak{O} = \mathfrak{P}$.

28. If $K = Q(\sqrt{-5})$, factor $3\mathfrak{O}$, $7\mathfrak{O}$, $21\mathfrak{O}$, $(4 + \sqrt{-5})\mathfrak{O}$, $(4 - \sqrt{-5})\mathfrak{O}$, $(1 + 2\sqrt{-5})\mathfrak{O}$, and $(1 - 2\sqrt{-5})\mathfrak{O}$ into products of prime ideals of $\mathfrak{O}$.

29. Determine the units and primes of the ring of integers of $Q(i)$ and show that this ring is a UFD.

30. Show that the units of the ring of integers of $Q(\sqrt{2})$ are $\pm(1 + \sqrt{2})^n$, $n = 0, \pm 1, \pm 2, \ldots$. This shows that the ring of integers of an algebraic number field may have infinitely many units.

31. Determine the different of K/Q.

32. Let $\mathfrak{P}$ be a prime ideal of $\mathfrak{O}$. Determine the decomposition group and field, inertia group and field, and the ramification groups and fields of $\mathfrak{P}$.

Miscellaneous exercises on algebraic number fields

33. Let $K = k(\zeta)$, where ζ is a primitive pth root of unity, p being an odd rational prime. Then $d(1, \zeta, \ldots, \zeta^{p-2}) = (-1)^{(p-1)/2} p^{p-2}$ (see Exercise 2.28).
   (a) If $a = 1 - \zeta$, show that $a\mathfrak{O}$ is a prime ideal of $\mathfrak{O}$, $N_{K/Q}(a\mathfrak{O}) = pJ$, and $p\mathfrak{O} = (a\mathfrak{O})^{p-1}$.
   (b) Show that $1, \zeta, \ldots \zeta^{p-2}$ is a minimal basis of K/Q, and draw the proper conclusions concerning ramification of prime ideals of J in $\mathfrak{O}$.

34. If K is the field of Exercise 33, determine the different of K/Q.

35. Let K be the field of Exercise 33, and let $\mathfrak{P}$ be a prime ideal of $\mathfrak{O}$. Determine the decomposition group and field, the inertia group and field, and the ramification groups and fields of $\mathfrak{P}$.

36. Let K be an algebraic number field.
   (a) Let $a_1, \ldots, a_n$ be a basis of K/Q with $a_i \in \mathfrak{O}$ for $i = 1, \ldots, n$. Show that if $d(a_1, \ldots, a_n)$ is square-free then $a_1, \ldots, a_n$ is a minimal basis of K/Q.
   (b) Theorem of Stickelberger. Show that $\Delta \equiv 0$ or $1 \pmod 4$.

37. Let $K = Q(a)$ where $\mathrm{Irr}\,(Q,a) = x^3 - x - 1$. Show that $\Delta = -23$ and that $1, a, a^2$ is a minimal basis of K/Q. Factor $p\mathfrak{O}$ in $\mathfrak{O}$ for several small rational primes p.

Proof of Theorem 19 of Chapter 2

We shall show that for $n \geqslant 5$, S_n is not a solvable group. The elements of S_n are permutations on some set of n symbols. We shall denote these symbols by $a, b, c, \ldots$. If S_n is solvable then A_n, the alternating subgroup of S_n, is solvable by Proposition 2 of Section 9 of Chapter 2. Hence it is sufficient to show that if $n \geqslant 5$ then A_n is not solvable. We note that for $n \geqslant 4$, A_n is not Abelian, for both (abc) and (abd) are in A_n and we have $(abc)(abd) = (ac)(bd)$ and $(abd)(abc) = (ad)(bc)$.

If $\sigma \in S_n$ then we can always write σ in the form $(ab)(ac) \cdots (am)$. If $\sigma \in A_n$ the number of these factors (called transpositions) is even. We note that each transposition is its own inverse, so that we have

$$(ac)(ad) = (ac)(ab)(ab)(ad) = (abc)(abd)^2.$$

Therefore, A_n is generated by the set of elements of S_n of the form (abc) where a and b are fixed symbols and c runs through the $n - 2$ symbols other than a and b.

Now let H be a normal subgroup of A_n and suppose that H contains an element of the form (abc). Then $(abc)^2 = (acb) \in H$. If $\sigma = (ab)(cd)$ then $\sigma^{-1} = \sigma$ and $\sigma(acb)\sigma^{-1} = (ab)(cd)(acb)(ab)(cd) = (abd) \in H$ for all symbols d. Therefore, $H = A_n$.

Suppose that H is an arbitrary normal subgroup of A_n other than the identity subgroup. Choose an element $\sigma \in H$, with $\sigma \neq 1$, such that no element of H other than the identity leaves fixed more symbols than does σ. If we write $\sigma = (ab)(ac) \cdots (am)$ then there must be at least two transpositions present. Since $\sigma \neq 1$, it moves at least two symbols and we may assume that it moves a and b. We now consider two cases. We may have $\sigma = (ab)(ac) = (acb)$, in which case we conclude that $H = A_n$. Otherwise we have $\sigma = (ab)(ac)(ad)(ae) \cdots$ (at least four transpositions). It is no loss of generality to assume that no two consecutive transpositions are the same. However, we might have $b = d$, in which case we have $\sigma = (ae)(cb) \cdots$. If $b \neq d$ then $\sigma = (adcb)(ae) \ldots$. If we rename our symbols we see that we may assume that either $\sigma = (abc \cdots) \cdots$ or $\sigma = (ab)(cd) \cdots$: in the second case b and d may be the same. In any case, if we assume $n \geqslant 5$, there is a symbol e that is different from a, b, c, and d. Then $\tau = (cde) \in A_n$ and $\tau^{-1} = (ced)$. Therefore either $\tau\sigma\tau^{-1} = (abd \cdots) \cdots$ or $\tau\sigma\tau^{-1} = (ab)(de) \cdots$ is in H. In both cases, $\sigma \neq \tau\sigma\tau^{-1}$ and so $\sigma^{-1}\tau\sigma\tau^{-1}$ belongs to H and is not the identity element of H. However, $\sigma^{-1}\tau\sigma\tau^{-1}$ leaves fixed more symbols than σ does, which contradicts our choice of σ. Therefore, this case is impossible and we must have $H = A_n$.

Example of the Galois Group of an Infinite Extension

Let $\mathfrak{o}^+$ be the additive group of the ring of p-adic integers. We leave it as an exercise to show that $\mathfrak{o}^+$ is a compact Hausdorff group (with respect to the topology induced by the p-adic valuation). For each $n \geqslant 1$, $p^n \mathfrak{o}^+$ is a closed subgroup of $\mathfrak{o}^+$ and $\bigcap_{n=1}^{\infty} p^n \mathfrak{o}^n = 0$. Let N be the set of positive integers ordered as usual. If $m \leqslant n$ we define $\phi_{nm} \colon \mathfrak{o}^+/p^n\mathfrak{o}^+ \to \mathfrak{o}^+/p^m\mathfrak{o}^+$ by setting $\phi_{nm}(a + p^n\mathfrak{o}^+) = a + p^m\mathfrak{o}^+$. Then $(N, \{\mathfrak{o}^+/p^n\mathfrak{o}^+\}, \{\phi_{nm}\})$ is an inverse system of groups and

$$\mathfrak{o}^+ \cong \varprojlim \mathfrak{o}^+/p^n\mathfrak{o}^+$$

by the corollary to Theorem 23 in Chapter 2.

It is easily seen that for $n \geqslant 1$, $\mathfrak{o}^+/p^n\mathfrak{o}^+$ is isomorphic to $J/(p^n)$, the cyclic group of order p^n, and therefore we have in a natural way

$$\mathfrak{o}^+ \cong \varprojlim J/(p^n).$$

Here the inverse system of the groups in question is $(N, \{J/(p^n)\}, \{\psi_{mn}\})$ where $\psi_{mn} \colon J/(p^n) \to J/(p^m)$ is given by $\psi_{mn}(a + (p^n)) = a + (p^m)$ for $m \leqslant n$, and the groups $J/(p^n)$ have the discrete topology.

Let k be a field and let C be an algebraic closure of k. Assume that for each non-negative integer n, k has a cyclic extension K_n in C with $[K_n \colon k] = p^n$ such that $K_n \subseteq K_{n+1}$ for $n \geqslant 0$. A finite field has such a tower of cyclic extensions. Let $K = \bigcup_{n=0}^{\infty} K_n$. Then K is certainly a separable extension of k, and it follows from Theorem 16a of Chapter 1 that K/k is normal. Let $\mathfrak{F}$ be the family of all finite normal extensions of k in K. If $L \in \mathfrak{F}$ then $L = k(a)$ for some $a \in K$. If $a \in K_r$ then $L \subseteq K_r$ and $G(K_r/L)$ is the unique subgroup of $G(K_r/k)$ of order, say, p^s. Hence L is the unique subfield of K_r with $[L \colon k] = p^{r-s}$. Therefore, $L = K_{r-s}$. Thus $\mathfrak{F} = \{K_n \mid n \geqslant 0\}$. If we combine these remarks with Theorem 24 of Chapter 2 we have $G(K/k) \cong \mathfrak{o}^+$, and the isomorphism is topological.

Bibliography

The following is a list of books and lecture notes which deal with topics we have considered. Many of these works will carry the reader far past the point we have reached in the final pages of this book. We have roughly classified these books and lecture notes by topics. However, there is some overlap between the various classes. For example, the book by O'Meara [43] contains an excellent introduction to valuation theory, while the notes by Artin [19] contain a good deal of material on algebraic functions.

Books and lecture notes which contain an exposition of general field theory including Galois theory.

1. ADAMSON, IAIN T. *Introduction to Field Theory.* New York: Interscience Publishers, 1965.
2. ALBERT, A. A. *Modern Higher Algebra.* Chicago: University of Chicago Press, 1937.
3. ALBERT, A. A. *Fundamental Concepts of Higher Algebra.* Chicago: University of Chicago Press, 1956.
4. ARTIN, E. *Galois Theory,* 2d Ed. University of Notre Dame, 1946.
5. ARTIN, E. Selected Topics in Modern Algebra, Lecture Notes. University of North Carolina, 1954.
6. ARTIN, E. Modern Higher Algebra; Galois Theory. Lecture Notes, New York University, 1955.
7. BOURBAKI, N. *Algèbre* (Chapitres IV et V). Paris: Hermann, 1950.
8. HASSE, H. *Higher Algebra.* New York: Ungar Publishing Co., 1954.
9. HAUPT, O. *Einführung in die Algebra, Zweiter Teil.* Leipzig: Akademische Verlagsgesellschaft, 1954.
10. N. HERSTEIN, I. *Topics in Algebra.* New York: Blaisdell Publishing Co., 1964.
11. JACOBSON, N. *Lectures in Abstract Algebra, Volume III.* Princeton: D. van Nostrand Co., Inc., 1964.
12. LUGOWSKI, H. and H. J. WEINERT. *Grundzüge der Algebra, Teil III.* Leipzig: B. G. Teubner Verlag, 1960.
13. POSTNIKOV, M. M. *Foundations of Galois Theory* (Translated by L. F. Boron). Groningen: P. Noordhoff N. V., 1962.
14. RAMANATHAN, K. G. *Lectures on the Algebraic Theory of Fields,* Lecture Notes, Tata Institute of Fundamental Research, 1954.

15. STEINITZ, E. *Algebraische Theorie der Körper.* New York: Chelsea Publishing Co., 1950. (First published in 1910.)
16. TSCHEBOTOROW, N. *Grundzüge der Galois'schen Theorie* (Übersetzt und Bearbeitet von H. Schwerdtfeger). Groningen: P. Noordhoff N.V., 1950.
17. VAN DER WAERDEN, B. L. *Modern Algebra, Volume I.* New York: Ungar Publishing Co., 1949.
18. ZARISKI, O. and P. SAMUEL. *Commutative Algebra, Volumes I and II.* Princeton: D. van Nostrand Co., Inc., 1958 and 1960.

Books and lectures note devoted to valuation theory, algebraic number theory, or class field theory.

19. ARTIN, E. *Algebraic Numbers and Algebraic Functions,* Lecture Notes, New York University, 1951.
20. ARTIN, E. *Theory of Algebraic Numbers.* Göttingen, 1959.
21. ARTIN, E. and J. TATE. *Class Field Theory.* Harvard University, 1961.
22. BACHMAN, G. *Introduction to p-adic Numbers and Valuation Theory.* New York: Academic Press, Inc., 1964.
23. BOURBAKI, N. *Algèbre Commutative* (Chapitres 5 et 6). Paris: Hermann, 1964.
24. CHEVALLEY, C. *Class Field Theory.* Nagoya University, 1954.
25. GÁL, I. S. *Lectures on Number Theory.* Minneapolis, 1961.
26. HANCOCK, H. *Foundations of the Theory of Algebraic Numbers, Volumes I and II.* New York: The Macmillan Co., 1931 and 1932.
27. HASSE, H. *Zahlentheorie,* 2d Ed. Berlin: Akademie-Verlag, 1963.
28. HECKE, E. *Vorlesungen über die Theorie der algebraischen Zahlen.* Leipzig: Akademischer Verlag, 1923.
29. HOLZER, L. *Zahlentheorie, Teil I und Teil II.* Leipzig: B. G. Teubner Verlag, 1959.
30. LANDAU, E. *Vorlesungen über Zahlentheorie, Dritter Band. Aus der algebraischen Zahlentheorie und über die Fermatsche Vermutung.* New York: Chelsea Publishing Co., 1947. (First published in 1927.)
31. LANG, S. *Algebraic Numbers.* Reading: Addison-Wesley Publishing Co., 1964.
32. MANN, H. B. *Introduction to Algebraic Number Theory.* Columbus: Ohio State University Press, 1955.
33. POLLARD, H. *The Theory of Algebraic Numbers.* Buffalo: Mathematical Association of America, 1950.
34. SCHILLING, O. F. G. *The Theory of Valuations.* New York: American Mathematical Society, 1950.
35. SERRE, J. P. *Corps Locaux.* Paris: Hermann, 1963.
36. SOMMER, J. *Vorlesungen über Zahlentheorie.* Leipzig: B. G. Teubner Verlag, 1907.
37. WEISS, E. *Algebraic Number Theory.* New York: McGraw-Hill Book Co., Inc., 1963.
38. WEYL, H. *Algebraic Theory of Numbers.* Princeton: Princeton University Press, 1940.

Books and lecture notes containing various applications of field theory and valuation theory.

39. ARTIN, E. *Elements of Algebraic Geometry, Lecture Notes,* New York University, 1955.
40. CHEVALLEY, C. *Introduction to the Theory of Algebraic Functions of One Variable.* New York: American Mathematical Society, 1951.
41. LANG, S. *Introduction to Algebraic Geometry.* New York: Interscience Publishers, 1959.
42. LANG, S. *Diophantine Geometry.* New York: Interscience Publishers, 1962.
43. O'MEARA, O. T. *Introduction to Quadratic Forms.* New York: Academic Press, Inc., 1963.
44. WEIL, A. *Foundations of Algebraic Geometry.* New York: American Mathematical Society, 1946.

Miscellaneous references

45. BOURBAKI, N. *Topologie Générale* (Chapitres I et II). Paris: Hermann, 1951.
46. BOURBAKI, N. *Topologie Générale* (Chapitres III et IV). Paris: Hermann, 1951.
47. KELLEY, J. L. *General Topology.* Princeton: D. van Nostrand Co., Inc., 1955.
48. RADEMACHER, H. *Lectures on Elementary Number Theory.* New York: Blaisdell Publishing Co., 1964.

Special reference

Anyone who professes an interest in the theory of fields, and especially those who wish to study algebraic number theory, should at least browse through the monumental work known as Hilbert's Zahlbericht.

49. D. HILBERT, *Bericht über die Theorie der algebraischen Zahlkörper, Jahresbericht der Deutschen Mathematiker-Vereinigung,* vol. 4, 1894–1895, pp. 175–546.

This paper also appears in *Hilbert's Gesammelte Abhandlungen, Erster Band,* Berlin: Springer Verlag, 1932, pp. 63–363. A translation of this paper into French appears as *"Theorie des corps de nombres algebriques" Annales de la Faculté des Sciences de l'Université de Toulouse,* vol. 1, 1909, pp. 257–329; vol. 2, 1910, pp. 225–456; vol. 3, 1911, pp. 1–62n. This translation contains a number of errors and must be read with care.

INDEX

A CATALOG OF SELECTED
DOVER BOOKS
IN SCIENCE AND MATHEMATICS

A CATALOG OF SELECTED
DOVER BOOKS
IN SCIENCE AND MATHEMATICS

QUALITATIVE THEORY OF DIFFERENTIAL EQUATIONS, V.V. Nemytskii and V.V. Stepanov. Classic graduate-level text by two prominent Soviet mathematicians covers classical differential equations as well as topological dynamics and erqodic theory. Bibliographies. 523pp. 5⅜ × 8½. 65954-2 Pa. $10.95

MATRICES AND LINEAR ALGEBRA, Hans Schneider and George Phillip Barker. Basic textbook covers theory of matrices and its applications to systems of linear equations and related topics such as determinants, eigenvalues and differential equations. Numerous exercises. 432pp. 5⅜ × 8½. 66014-1 Pa. $8.95

QUANTUM THEORY, David Bohm. This advanced undergraduate-level text presents the quantum theory in terms of qualitative and imaginative concepts, followed by specific applications worked out in mathematical detail. Preface. Index. 655pp. 5⅜ × 8½. 65969-0 Pa. $10.95

ATOMIC PHYSICS (8th edition), Max Born. Nobel laureate's lucid treatment of kinetic theory of gases, elementary particles, nuclear atom, wave-corpuscles, atomic structure and spectral lines, much more. Over 40 appendices, bibliography. 495pp. 5⅜ × 8½. 65984-4 Pa. $11.95

ELECTRONIC STRUCTURE AND THE PROPERTIES OF SOLIDS: The Physics of the Chemical Bond, Walter A. Harrison. Innovative text offers basic understanding of the electronic structure of covalent and ionic solids, simple metals, transition metals and their compounds. Problems. 1980 edition. 582pp. 6⅛ × 9¼. 66021-4 Pa. $14.95

BOUNDARY VALUE PROBLEMS OF HEAT CONDUCTION, M. Necati Özisik. Systematic, comprehensive treatment of modern mathematical methods of solving problems in heat conduction and diffusion. Numerous examples and problems. Selected references. Appendices. 505pp. 5⅜ × 8½. 65990-9 Pa. $11.95

A SHORT HISTORY OF CHEMISTRY (3rd edition), J.R. Partington. Classic exposition explores origins of chemistry, alchemy, early medical chemistry, nature of atmosphere, theory of valency, laws and structure of atomic theory, much more. 428pp. 5⅜ × 8½. (Available in U.S. only) 65977-1 Pa. $10.95

A HISTORY OF ASTRONOMY, A. Pannekoek. Well-balanced, carefully reasoned study covers such topics as Ptolemaic theory, work of Copernicus, Kepler, Newton, Eddington's work on stars, much more. Illustrated. References. 521pp. 5⅜ × 8½. 65994-1 Pa. $11.95

PRINCIPLES OF METEOROLOGICAL ANALYSIS, Walter J. Saucier. Highly respected, abundantly illustrated classic reviews atmospheric variables, hydrostatics, static stability, various analyses (scalar, cross-section, isobaric, isentropic, more). For intermediate meteorology students. 454pp. 6⅛ × 9¼. 65979-8 Pa. $12.95

RELATIVITY, THERMODYNAMICS AND COSMOLOGY, Richard C. Tolman. Landmark study extends thermodynamics to special, general relativity; also applications of relativistic mechanics, thermodynamics to cosmological models. 501pp. 5⅜ × 8½. 65383-8 Pa. $11.95

APPLIED ANALYSIS, Cornelius Lanczos. Classic work on analysis and design of finite processes for approximating solution of analytical problems. Algebraic equations, matrices, harmonic analysis, quadrature methods, much more. 559pp. 5⅜ × 8½. 65656-X Pa. $11.95

SPECIAL RELATIVITY FOR PHYSICISTS, G. Stephenson and C.W. Kilmister. Concise elegant account for nonspecialists. Lorentz transformation, optical and dynamical applications, more. Bibliography. 108pp. 5⅜ × 8½. 65519-9 Pa. $3.95

INTRODUCTION TO ANALYSIS, Maxwell Rosenlicht. Unusually clear, accessible coverage of set theory, real number system, metric spaces, continuous functions, Riemann integration, multiple integrals, more. Wide range of problems. Undergraduate level. Bibliography. 254pp. 5⅜ × 8½. 65038-3 Pa. $7.00

INTRODUCTION TO QUANTUM MECHANICS With Applications to Chemistry, Linus Pauling & E. Bright Wilson, Jr. Classic undergraduate text by Nobel Prize winner applies quantum mechanics to chemical and physical problems. Numerous tables and figures enhance the text. Chapter bibliographies. Appendices. Index. 468pp. 5⅜ × 8½. 64871-0 Pa. $9.95

ASYMPTOTIC EXPANSIONS OF INTEGRALS, Norman Bleistein & Richard A. Handelsman. Best introduction to important field with applications in a variety of scientific disciplines. New preface. Problems. Diagrams. Tables. Bibliography. Index. 448pp. 5⅜ × 8½. 65082-0 Pa. $10.95

MATHEMATICS APPLIED TO CONTINUUM MECHANICS, Lee A. Segel. Analyzes models of fluid flow and solid deformation. For upper-level math, science and engineering students. 608pp. 5⅜ × 8½. 65369-2 Pa. $12.95

ELEMENTS OF REAL ANALYSIS, David A. Sprecher. Classic text covers fundamental concepts, real number system, point sets, functions of a real variable, Fourier series, much more. Over 500 exercises. 352pp. 5⅜ × 8½. 65385-4 Pa. $8.95

PHYSICAL PRINCIPLES OF THE QUANTUM THEORY, Werner Heisenberg. Nobel Laureate discusses quantum theory, uncertainty, wave mechanics, work of Dirac, Schroedinger, Compton, Wilson, Einstein, etc. 184pp. 5⅜ × 8½. 60113-7 Pa. $4.95

INTRODUCTORY REAL ANALYSIS, A.N. Kolmogorov, S.V. Fomin. Translated by Richard A. Silverman. Self-contained, evenly paced introduction to real and functional analysis. Some 350 problems. 403pp. 5⅜ × 8½. 61226-0 Pa. $7.95

PROBLEMS AND SOLUTIONS IN QUANTUM CHEMISTRY AND PHYSICS, Charles S. Johnson, Jr. and Lee G. Pedersen. Unusually varied problems, detailed solutions in coverage of quantum mechanics, wave mechanics, angular momentum, molecular spectroscopy, scattering theory, more. 280 problems plus 139 supplementary exercises. 430pp. 6½ × 9¼. 65236-X Pa. $10.95

ASYMPTOTIC METHODS IN ANALYSIS, N.G. de Bruijn. An inexpensive, comprehensive guide to asymptotic methods—the pioneering work that teaches by explaining worked examples in detail. Index. 224pp. 5⅜ × 8½.   64221-6 Pa. $5.95

OPTICAL RESONANCE AND TWO-LEVEL ATOMS, L. Allen and J.H. Eberly. Clear, comprehensive introduction to basic principles behind all quantum optical resonance phenomena. 53 illustrations. Preface. Index. 256pp. 5⅜ × 8½.
65533-4 Pa. $6.95

COMPLEX VARIABLES, Francis J. Flanigan. Unusual approach, delaying complex algebra till harmonic functions have been analyzed from real variable viewpoint. Includes problems with answers. 364pp. 5⅜ × 8½.   61388-7 Pa. $7.95

ATOMIC SPECTRA AND ATOMIC STRUCTURE, Gerhard Herzberg. One of best introductions; especially for specialist in other fields. Treatment is physical rather than mathematical. 80 illustrations. 257pp. 5⅜ × 8½.   60115-3 Pa. $4.95

APPLIED COMPLEX VARIABLES, John W. Dettman. Step-by-step coverage of fundamentals of analytic function theory—plus lucid exposition of 5 important applications: Potential Theory; Ordinary Differential Equations; Fourier Transforms; Laplace Transforms; Asymptotic Expansions. 66 figures. Exercises at chapter ends. 512pp. 5⅜ × 8½.   64670-X Pa. $10.95

ULTRASONIC ABSORPTION: An Introduction to the Theory of Sound Absorption and Dispersion in Gases, Liquids and Solids, A.B. Bhatia. Standard reference in the field provides a clear, systematically organized introductory review of fundamental concepts for advanced graduate students, research workers. Numerous diagrams. Bibliography. 440pp. 5⅜ × 8½.   64917-2 Pa. $8.95

UNBOUNDED LINEAR OPERATORS: Theory and Applications, Seymour Goldberg. Classic presents systematic treatment of the theory of unbounded linear operators in normed linear spaces with applications to differential equations. Bibliography. 199pp. 5⅜ × 8½.   64830-3 Pa. $7.00

LIGHT SCATTERING BY SMALL PARTICLES, H.C. van de Hulst. Comprehensive treatment including full range of useful approximation methods for researchers in chemistry, meteorology and astronomy. 44 illustrations. 470pp. 5⅜ × 8½.   64228-3 Pa. $9.95

CONFORMAL MAPPING ON RIEMANN SURFACES, Harvey Cohn. Lucid, insightful book presents ideal coverage of subject. 334 exercises make book perfect for self-study. 55 figures. 352pp. 5⅜ × 8¼.   64025-6 Pa. $8.95

OPTICKS, Sir Isaac Newton. Newton's own experiments with spectroscopy, colors, lenses, reflection, refraction, etc., in language the layman can follow. Foreword by Albert Einstein. 532pp. 5⅜ × 8½.   60205-2 Pa. $8.95

GENERALIZED INTEGRAL TRANSFORMATIONS, A.H. Zemanian. Graduate-level study of recent generalizations of the Laplace, Mellin, Hankel, K. Weierstrass, convolution and other simple transformations. Bibliography. 320pp. 5⅜ × 8½.   65375-7 Pa. $7.95

THE ELECTROMAGNETIC FIELD, Albert Shadowitz. Comprehensive undergraduate text covers basics of electric and magnetic fields, builds up to electromagnetic theory. Also related topics, including relativity. Over 900 problems. 768pp. 5⅜ × 8¼. 65660-8 Pa. $15.95

FOURIER SERIES, Georgi P. Tolstov. Translated by Richard A. Silverman. A valuable addition to the literature on the subject, moving clearly from subject to subject and theorem to theorem. 107 problems, answers. 336pp. 5⅜ × 8½. 63317-9 Pa. $7.95

THEORY OF ELECTROMAGNETIC WAVE PROPAGATION, Charles Herach Papas. Graduate-level study discusses the Maxwell field equations, radiation from wire antennas, the Doppler effect and more. xiii + 244pp. 5⅜ × 8½. 65678-0 Pa. $6.95

DISTRIBUTION THEORY AND TRANSFORM ANALYSIS: An Introduction to Generalized Functions, with Applications, A.H. Zemanian. Provides basics of distribution theory, describes generalized Fourier and Laplace transformations. Numerous problems. 384pp. 5⅜ × 8½. 65479-6 Pa. $8.95

THE PHYSICS OF WAVES, William C. Elmore and Mark A. Heald. Unique overview of classical wave theory. Acoustics, optics, electromagnetic radiation, more. Ideal as classroom text or for self-study. Problems. 477pp. 5⅜ × 8½. 64926-1 Pa. $10.95

CALCULUS OF VARIATIONS WITH APPLICATIONS, George M. Ewing. Applications-oriented introduction to variational theory develops insight and promotes understanding of specialized books, research papers. Suitable for advanced undergraduate/graduate students as primary, supplementary text. 352pp. 5⅜ × 8½. 64856-7 Pa. $8.50

A TREATISE ON ELECTRICITY AND MAGNETISM, James Clerk Maxwell. Important foundation work of modern physics. Brings to final form Maxwell's theory of electromagnetism and rigorously derives his general equations of field theory. 1,084pp. 5⅜ × 8½. 60636-8, 60637-6 Pa., Two-vol. set $19.00

AN INTRODUCTION TO THE CALCULUS OF VARIATIONS, Charles Fox. Graduate-level text covers variations of an integral, isoperimetrical problems, least action, special relativity, approximations, more. References. 279pp. 5⅜ × 8½. 65499-0 Pa. $6.95

HYDRODYNAMIC AND HYDROMAGNETIC STABILITY, S. Chandrasekhar. Lucid examination of the Rayleigh-Benard problem; clear coverage of the theory of instabilities causing convection. 704pp. 5⅜ × 8¼. 64071-X Pa. $12.95

CALCULUS OF VARIATIONS, Robert Weinstock. Basic introduction covering isoperimetric problems, theory of elasticity, quantum mechanics, electrostatics, etc. Exercises throughout. 326pp. 5⅜ × 8½. 63069-2 Pa. $7.95

DYNAMICS OF FLUIDS IN POROUS MEDIA, Jacob Bear. For advanced students of ground water hydrology, soil mechanics and physics, drainage and irrigation engineering and more. 335 illustrations. Exercises, with answers. 784pp. 6⅛ × 9¼. 65675-6 Pa. $19.95

NUMERICAL METHODS FOR SCIENTISTS AND ENGINEERS, Richard Hamming. Classic text stresses frequency approach in coverage of algorithms, polynomial approximation, Fourier approximation, exponential approximation, other topics. Revised and enlarged 2nd edition. 721pp. 5⅜ × 8½.
65241-6 Pa. $14.95

THEORETICAL SOLID STATE PHYSICS, Vol. I: Perfect Lattices in Equilibrium; Vol. II: Non-Equilibrium and Disorder, William Jones and Norman H. March. Monumental reference work covers fundamental theory of equilibrium properties of perfect crystalline solids, non-equilibrium properties, defects and disordered systems. Appendices. Problems. Preface. Diagrams. Index. Bibliography. Total of 1,301pp. 5⅜ × 8½. Two volumes. Vol. I 65015-4 Pa. $12.95
Vol. II 65016-2 Pa. $12.95

OPTIMIZATION THEORY WITH APPLICATIONS, Donald A. Pierre. Broad-spectrum approach to important topic. Classical theory of minima and maxima, calculus of variations, simplex technique and linear programming, more. Many problems, examples. 640pp. 5⅜ × 8½. 65205-X Pa. $12.95

THE MODERN THEORY OF SOLIDS, Frederick Seitz. First inexpensive edition of classic work on theory of ionic crystals, free-electron theory of metals and semiconductors, molecular binding, much more. 736pp. 5⅜ × 8½.
65482-6 Pa. $14.95

ESSAYS ON THE THEORY OF NUMBERS, Richard Dedekind. Two classic essays by great German mathematician: on the theory of irrational numbers; and on transfinite numbers and properties of natural numbers. 115pp. 5⅜ × 8½.
21010-3 Pa. $4.95

THE FUNCTIONS OF MATHEMATICAL PHYSICS, Harry Hochstadt. Comprehensive treatment of orthogonal polynomials, hypergeometric functions, Hill's equation, much more. Bibliography. Index. 322pp. 5⅜ × 8½. 65214-9 Pa. $8.95

NUMBER THEORY AND ITS HISTORY, Oystein Ore. Unusually clear, accessible introduction covers counting, properties of numbers, prime numbers, much more. Bibliography. 380pp. 5⅜ × 8½. 65620-9 Pa. $8.95

THE VARIATIONAL PRINCIPLES OF MECHANICS, Cornelius Lanczos. Graduate level coverage of calculus of variations, equations of motion, relativistic mechanics, more. First inexpensive paperbound edition of classic treatise. Index. Bibliography. 418pp. 5⅜ × 8½. 65067-7 Pa. $10.95

MATHEMATICAL TABLES AND FORMULAS, Robert D. Carmichael and Edwin R. Smith. Logarithms, sines, tangents, trig functions, powers, roots, reciprocals, exponential and hyperbolic functions, formulas and theorems. 269pp. 5⅜ × 8½. 60111-0 Pa. $5.95

THEORETICAL PHYSICS, Georg Joos, with Ira M. Freeman. Classic overview covers essential math, mechanics, electromagnetic theory, thermodynamics, quantum mechanics, nuclear physics, other topics. First paperback edition. xxiii + 885pp. 5⅜ × 8½. 65227-0 Pa. $17.95

HANDBOOK OF MATHEMATICAL FUNCTIONS WITH FORMULAS, GRAPHS, AND MATHEMATICAL TABLES, edited by Milton Abramowitz and Irene A. Stegun. Vast compendium: 29 sets of tables, some to as high as 20 places. 1,046pp. 8 × 10½. 61272-4 Pa. $21.95

MATHEMATICAL METHODS IN PHYSICS AND ENGINEERING, John W. Dettman. Algebraically based approach to vectors, mapping, diffraction, other topics in applied math. Also generalized functions, analytic function theory, more. Exercises. 448pp. 5⅜ × 8¼. 65649-7 Pa. $8.95

A SURVEY OF NUMERICAL MATHEMATICS, David M. Young and Robert Todd Gregory. Broad self-contained coverage of computer-oriented numerical algorithms for solving various types of mathematical problems in linear algebra, ordinary and partial, differential equations, much more. Exercises. Total of 1,248pp. 5⅜ × 8½. Two volumes. Vol. I 65691-8 Pa. $13.95
Vol. II 65692-6 Pa. $13.95

TENSOR ANALYSIS FOR PHYSICISTS, J.A. Schouten. Concise exposition of the mathematical basis of tensor analysis, integrated with well-chosen physical examples of the theory. Exercises. Index. Bibliography. 289pp. 5⅜ × 8½. 65582-2 Pa. $7.95

INTRODUCTION TO NUMERICAL ANALYSIS (2nd Edition), F.B. Hildebrand. Classic, fundamental treatment covers computation, approximation, interpolation, numerical differentiation and integration, other topics. 150 new problems. 669pp. 5⅜ × 8½. 65363-3 Pa. $13.95

INVESTIGATIONS ON THE THEORY OF THE BROWNIAN MOVEMENT, Albert Einstein. Five papers (1905–8) investigating dynamics of Brownian motion and evolving elementary theory. Notes by R. Fürth. 122pp. 5⅜ × 8½. 60304-0 Pa. $3.95

NUMERICAL METHODS FOR SCIENTISTS AND ENGINEERS, Richard Hamming. Classic text stresses frequency approach in coverage of algorithms, polynomial approximation, Fourier approximation, exponential approximation, other topics. Revised and enlarged 2nd edition. 721pp. 5⅜ × 8½. 65241-6 Pa. $14.95

AN INTRODUCTION TO STATISTICAL THERMODYNAMICS, Terrell L. Hill. Excellent basic text offers wide-ranging coverage of quantum statistical mechanics, systems of interacting molecules, quantum statistics, more. 523pp. 5⅜ × 8½. 65242-4 Pa. $10.95

ELEMENTARY DIFFERENTIAL EQUATIONS, William Ted Martin and Eric Reissner. Exceptionally, clear comprehensive introduction at undergraduate level. Nature and origin of differential equations, differential equations of first, second and higher orders. Picard's Theorem, much more. Problems with solutions. 331pp. 5⅜ × 8½. 65024-3 Pa. $8.95

STATISTICAL PHYSICS, Gregory H. Wannier. Classic text combines thermodynamics, statistical mechanics and kinetic theory in one unified presentation of thermal physics. Problems with solutions. Bibliography. 532pp. 5⅜ × 8½. 65401-X Pa. $10.95

SPECIAL FUNCTIONS, N.N. Lebedev. Translated by Richard Silverman. Famous Russian work treating more important special functions, with applications to specific problems of physics and engineering. 38 figures. 308pp. 5⅜ × 8½.
60624-4 Pa. $6.95

OBSERVATIONAL ASTRONOMY FOR AMATEURS, J.B. Sidgwick. Mine of useful data for observation of sun, moon, planets, asteroids, aurorae, meteors, comets, variables, binaries, etc. 39 illustrations 384pp. 5⅜ × 8¼. (Available in U.S. only)
24033-9 Pa. $5.95

INTEGRAL EQUATIONS, F.G. Tricomi. Authoritative, well-written treatment of extremely useful mathematical tool with wide applications. Volterra Equations, Fredholm Equations, much more. Advanced undergraduate to graduate level. Exercises. Bibliography. 238pp. 5⅜ × 8½.
64828-1 Pa. $6.95

CELESTIAL OBJECTS FOR COMMON TELESCOPES, T.W. Webb. Inestimable aid for locating and identifying nearly 4,000 celestial objects. 77 illustrations. 645pp. 5⅜ × 8½.
20917-2, 20918-0 Pa., Two-vol. set $12.00

MODERN NONLINEAR EQUATIONS, Thomas L. Saaty. Emphasizes practical solution of problems; covers seven types of equations. ". . . a welcome contribution to the existing literature. . . ."—*Math Reviews.* 490pp. 5⅜ × 8½. 64232-1 Pa. $9.95

FUNDAMENTALS OF ASTRODYNAMICS, Roger Bate et al. Modern approach developed by U.S. Air Force Academy. Designed as a first course. Problems, exercises. Numerous illustrations. 455pp. 5⅜ × 8½.
60061-0 Pa. $8.95

INTRODUCTION TO LINEAR ALGEBRA AND DIFFERENTIAL EQUATIONS, John W. Dettman. Excellent text covers complex numbers, determinants, orthonormal bases, Laplace transforms, much more. Exercises with solutions. Undergraduate level. 416pp. 5⅜ × 8½.
65191-6 Pa. $8.95

INCOMPRESSIBLE AERODYNAMICS, edited by Bryan Thwaites. Covers theoretical and experimental treatment of the uniform flow of air and viscous fluids past two-dimensional aerofoils and three-dimensional wings; many other topics. 654pp. 5⅜ × 8½.
65465-6 Pa. $14.95

INTRODUCTION TO DIFFERENCE EQUATIONS, Samuel Goldberg. Exceptionally clear exposition of important discipline with applications to sociology, psychology, economics. Many illustrative examples; over 250 problems. 260pp. 5⅜ × 8½.
65084-7 Pa. $6.95

LAMINAR BOUNDARY LAYERS, edited by L. Rosenhead. Engineering classic covers steady boundary layers in two- and three-dimensional flow, unsteady boundary layers, stability, observational techniques, much more. 708pp. 5⅜ × 8½.
65646-2 Pa. $15.95

LECTURES ON CLASSICAL DIFFERENTIAL GEOMETRY, Second Edition, Dirk J. Struik. Excellent brief introduction covers curves, theory of surfaces, fundamental equations, geometry on a surface, conformal mapping, other topics. Problems. 240pp. 5⅜ × 8½.
65609-8 Pa. $6.95

ORDINARY DIFFERENTIAL EQUATIONS, Morris Tenenbaum and Harry Pollard. Exhaustive survey of ordinary differential equations for undergraduates in mathematics, engineering, science. Thorough analysis of theorems. Diagrams. Bibliography. Index. 818pp. 5⅜ × 8½.   64940-7 Pa. $15.95

STATISTICAL MECHANICS: Principles and Applications, Terrell L. Hill. Standard text covers fundamentals of statistical mechanics, applications to fluctuation theory, imperfect gases, distribution functions, more. 448pp. 5⅜ × 8½.
65390-0 Pa. $9.95

ORDINARY DIFFERENTIAL EQUATIONS AND STABILITY THEORY: An Introduction, David A. Sánchez. Brief, modern treatment. Linear equation, stability theory for autonomous and nonautonomous systems, etc. 164pp. 5⅜ × 8¼.
63828-6 Pa. $4.95

THIRTY YEARS THAT SHOOK PHYSICS: The Story of Quantum Theory, George Gamow. Lucid, accessible introduction to influential theory of energy and matter. Careful explanations of Dirac's anti-particles, Bohr's model of the atom, much more. 12 plates. Numerous drawings. 240pp. 5⅜ × 8½.   24895-X Pa. $5.95

ORDINARY DIFFERENTIAL EQUATIONS, I.G. Petrovski. Covers basic concepts, some differential equations and such aspects of the general theory as Euler lines, Arzel's theorem, Peano's existence theorem, Osgood's uniqueness theorem, more. 45 figures. Problems. Bibliography. Index. xi + 232pp. 5⅜ × 8½.
64683-1 Pa. $6.00

GREAT EXPERIMENTS IN PHYSICS: Firsthand Accounts from Galileo to Einstein, edited by Morris H. Shamos. 25 crucial discoveries: Newton's laws of motion, Chadwick's study of the neutron, Hertz on electromagnetic waves, more. Original accounts clearly annotated. 370pp. 5⅜ × 8½.   25346-5 Pa. $8.95

INTRODUCTION TO PARTIAL DIFFERENTIAL EQUATIONS WITH AP-PLICATIONS, E.C. Zachmanoglou and Dale W. Thoe. Essentials of partial differential equations applied to common problems in engineering and the physical sciences. Problems and answers. 416pp. 5⅜ × 8½.   65251-3 Pa. $9.95

BURNHAM'S CELESTIAL HANDBOOK, Robert Burnham, Jr. Thorough guide to the stars beyond our solar system. Exhaustive treatment. Alphabetical by constellation: Andromeda to Cetus in Vol. 1; Chamaeleon to Orion in Vol. 2; and Pavo to Vulpecula in Vol. 3. Hundreds of illustrations. Index in Vol. 3. 2,000pp. 6⅛ × 9¼.   23567-X, 23568-8, 23673-0 Pa., Three-vol. set $38.85

ASYMPTOTIC EXPANSIONS FOR ORDINARY DIFFERENTIAL EQUA-TIONS, Wolfgang Wasow. Outstanding text covers asymptotic power series, Jordan's canonical form, turning point problems, singular perturbations, much more. Problems. 384pp. 5⅜ × 8½.   65456-7 Pa. $8.95

AMATEUR ASTRONOMER'S HANDBOOK, J.B. Sidgwick. Timeless, comprehensive coverage of telescopes, mirrors, lenses, mountings, telescope drives, micrometers, spectroscopes, more. 189 illustrations. 576pp. 5⅜ × 8¼.
24034-7 Pa. $8.95

ROTARY-WING AERODYNAMICS, W.Z. Stepniewski. Clear, concise text covers aerodynamic phenomena of the rotor and offers guidelines for helicopter performance evaluation. Originally prepared for NASA. 537 figures. 640pp. 6⅛ × 9¼.
64647-5 Pa. $14.95

DIFFERENTIAL GEOMETRY, Heinrich W. Guggenheimer. Local differential geometry as an application of advanced calculus and linear algebra. Curvature, transformation groups, surfaces, more. Exercises. 62 figures. 378pp. 5⅜ × 8½.
63433-7 Pa. $7.95

INTRODUCTION TO SPACE DYNAMICS, William Tyrrell Thomson. Comprehensive, classic introduction to space-flight engineering for advanced undergraduate and graduate students. Includes vector algebra, kinematics, transformation of coordinates. Bibliography. Index. 352pp. 5⅜ × 8½.   65113-4 Pa. $8.00

A SURVEY OF MINIMAL SURFACES, Robert Osserman. Up-to-date, in-depth discussion of the field for advanced students. Corrected and enlarged edition covers new developments. Includes numerous problems. 192pp. 5⅜ × 8½.
64998-9 Pa. $8.00

ANALYTICAL MECHANICS OF GEARS, Earle Buckingham. Indispensable reference for modern gear manufacture covers conjugate gear-tooth action, gear-tooth profiles of various gears, many other topics. 263 figures. 102 tables. 546pp. 5⅜ × 8½.   65712-4 Pa. $11.95

SET THEORY AND LOGIC, Robert R. Stoll. Lucid introduction to unified theory of mathematical concepts. Set theory and logic seen as tools for conceptual understanding of real number system. 496pp. 5⅜ × 8¼.   63829-4 Pa. $8.95

A HISTORY OF MECHANICS, René Dugas. Monumental study of mechanical principles from antiquity to quantum mechanics. Contributions of ancient Greeks, Galileo, Leonardo, Kepler, Lagrange, many others. 671pp. 5⅜ × 8½.
65632-2 Pa. $14.95

FAMOUS PROBLEMS OF GEOMETRY AND HOW TO SOLVE THEM, Benjamin Bold. Squaring the circle, trisecting the angle, duplicating the cube: learn their history, why they are impossible to solve, then solve them yourself. 128pp. 5⅜ × 8½.   24297-8 Pa. $3.95

MECHANICAL VIBRATIONS, J.P. Den Hartog. Classic textbook offers lucid explanations and illustrative models, applying theories of vibrations to a variety of practical industrial engineering problems. Numerous figures. 233 problems, solutions. Appendix. Index. Preface. 436pp. 5⅜ × 8½.   64785-4 Pa. $8.95

CURVATURE AND HOMOLOGY, Samuel I. Goldberg. Thorough treatment of specialized branch of differential geometry. Covers Riemannian manifolds, topology of differentiable manifolds, compact Lie groups, other topics. Exercises. 315pp. 5⅜ × 8½.   64314-X Pa. $6.95

HISTORY OF STRENGTH OF MATERIALS, Stephen P. Timoshenko. Excellent historical survey of the strength of materials with many references to the theories of elasticity and structure. 245 figures. 452pp. 5⅜ × 8½.   61187-6 Pa. $9.95

GEOMETRY OF COMPLEX NUMBERS, Hans Schwerdtfeger. Illuminating, widely praised book on analytic geometry of circles, the Moebius transformation, and two-dimensional non-Euclidean geometries. 200pp. 5⅜ × 8¼.

63830-8 Pa. $6.95

MECHANICS, J.P. Den Hartog. A classic introductory text or refresher. Hundreds of applications and design problems illuminate fundamentals of trusses, loaded beams and cables, etc. 334 answered problems. 462pp. 5⅜ × 8½. 60754-2 Pa. $8.95

TOPOLOGY, John G. Hocking and Gail S. Young. Superb one-year course in classical topology. Topological spaces and functions, point-set topology, much more. Examples and problems. Bibliography. Index. 384pp. 5⅜ × 8¼.

65676-4 Pa. $7.95

STRENGTH OF MATERIALS, J.P. Den Hartog. Full, clear treatment of basic material (tension, torsion, bending, etc.) plus advanced material on engineering methods, applications. 350 answered problems. 323pp. 5⅜ × 8½. 60755-0 Pa. $7.50

ELEMENTARY CONCEPTS OF TOPOLOGY, Paul Alexandroff. Elegant, intuitive approach to topology from set-theoretic topology to Betti groups; how concepts of topology are useful in math and physics. 25 figures. 57pp. 5⅜ × 8½.

60747-X Pa. $2.95

ADVANCED STRENGTH OF MATERIALS, J.P. Den Hartog. Superbly written advanced text covers torsion, rotating disks, membrane stresses in shells, much more. Many problems and answers. 388pp. 5⅜ × 8½. 65407-9 Pa. $8.95

COMPUTABILITY AND UNSOLVABILITY, Martin Davis. Classic graduate-level introduction to theory of computability, usually referred to as theory of recurrent functions. New preface and appendix. 288pp. 5⅜ × 8½. 61471-9 Pa. $6.95

GENERAL CHEMISTRY, Linus Pauling. Revised 3rd edition of classic first-year text by Nobel laureate. Atomic and molecular structure, quantum mechanics, statistical mechanics, thermodynamics correlated with descriptive chemistry. Problems. 992pp. 5⅜ × 8½. 65622-5 Pa. $18.95

AN INTRODUCTION TO MATRICES, SETS AND GROUPS FOR SCIENCE STUDENTS, G. Stephenson. Concise, readable text introduces sets, groups, and most importantly, matrices to undergraduate students of physics, chemistry, and engineering. Problems. 164pp. 5⅜ × 8½. 65077-4 Pa. $5.95

THE HISTORICAL BACKGROUND OF CHEMISTRY, Henry M. Leicester. Evolution of ideas, not individual biography. Concentrates on formulation of a coherent set of chemical laws. 260pp. 5⅜ × 8½. 61053-5 Pa. $6.00

THE PHILOSOPHY OF MATHEMATICS: An Introductory Essay, Stephan Körner. Surveys the views of Plato, Aristotle, Leibniz & Kant concerning propositions and theories of applied and pure mathematics. Introduction. Two appendices. Index. 198pp. 5⅜ × 8½. 25048-2 Pa. $5.95

THE DEVELOPMENT OF MODERN CHEMISTRY, Aaron J. Ihde. Authoritative history of chemistry from ancient Greek theory to 20th-century innovation. Covers major chemists and their discoveries. 209 illustrations. 14 tables. Bibliographies. Indices. Appendices. 851pp. 5⅜ × 8½. 64235-6 Pa. $15.95

THE FOUR-COLOR PROBLEM: Assaults and Conquest, Thomas L. Saaty and Paul G. Kainen. Engrossing, comprehensive account of the century-old combinatorial topological problem, its history and solution. Bibliographies. Index. 110 figures. 228pp. 5⅜ × 8½. 65092-8 Pa. $6.00

CATALYSIS IN CHEMISTRY AND ENZYMOLOGY, William P. Jencks. Exceptionally clear coverage of mechanisms for catalysis, forces in aqueous solution, carbonyl- and acyl-group reactions, practical kinetics, more. 864pp. 5⅜ × 8½. 65460-5 Pa. $18.95

PROBABILITY: An Introduction, Samuel Goldberg. Excellent basic text covers set theory, probability theory for finite sample spaces, binomial theorem, much more. 360 problems. Bibliographies. 322pp. 5⅜ × 8½. 65252-1 Pa. $7.95

LIGHTNING, Martin A. Uman. Revised, updated edition of classic work on the physics of lightning. Phenomena, terminology, measurement, photography, spectroscopy, thunder, more. Reviews recent research. Bibliography. Indices. 320pp. 5⅜ × 8¼. 64575-4 Pa. $7.95

PROBABILITY THEORY: A Concise Course, Y.A. Rozanov. Highly readable, self-contained introduction covers combination of events, dependent events, Bernoulli trials, etc. Translation by Richard Silverman. 148pp. 5⅜ × 8¼. 63544-9 Pa. $4.50

THE CEASELESS WIND: An Introduction to the Theory of Atmospheric Motion, John A. Dutton. Acclaimed text integrates disciplines of mathematics and physics for full understanding of dynamics of atmospheric motion. Over 400 problems. Index. 97 illustrations. 640pp. 6 × 9. 65096-0 Pa. $16.95

STATISTICS MANUAL, Edwin L. Crow, et al. Comprehensive, practical collection of classical and modern methods prepared by U.S. Naval Ordnance Test Station. Stress on use. Basics of statistics assumed. 288pp. 5⅜ × 8½. 60599-X Pa. $6.00

WIND WAVES: Their Generation and Propagation on the Ocean Surface, Blair Kinsman. Classic of oceanography offers detailed discussion of stochastic processes and power spectral analysis that revolutionized ocean wave theory. Rigorous, lucid. 676pp. 5⅜ × 8½. 64652-1 Pa. $14.95

STATISTICAL METHOD FROM THE VIEWPOINT OF QUALITY CONTROL, Walter A. Shewhart. Important text explains regulation of variables, uses of statistical control to achieve quality control in industry, agriculture, other areas. 192pp. 5⅜ × 8½. 65232-7 Pa. $6.00

THE INTERPRETATION OF GEOLOGICAL PHASE DIAGRAMS, Ernest G. Ehlers. Clear, concise text emphasizes diagrams of systems under fluid or containing pressure; also coverage of complex binary systems, hydrothermal melting, more. 288pp. 6½ × 9¼. 65389-7 Pa. $8.95

STATISTICAL ADJUSTMENT OF DATA, W. Edwards Deming. Introduction to basic concepts of statistics, curve fitting, least squares solution, conditions without parameter, conditions containing parameters. 26 exercises worked out. 271pp. 5⅜ × 8½. 64685-8 Pa. $7.95

DE RE METALLICA, Georgius Agricola. The famous Hoover translation of greatest treatise on technological chemistry, engineering, geology, mining of early modern times (1556). All 289 original woodcuts. 638pp. 6¾ × 11.
60006-8 Clothbd. $15.95

SOME THEORY OF SAMPLING, William Edwards Deming. Analysis of the problems, theory and design of sampling techniques for social scientists, industrial managers and others who find statistics increasingly important in their work. 61 tables. 90 figures. xvii + 602pp. 5⅜ × 8½. 64684-X Pa. $14.95

THE VARIOUS AND INGENIOUS MACHINES OF AGOSTINO RAMELLI: A Classic Sixteenth-Century Illustrated Treatise on Technology, Agostino Ramelli. One of the most widely known and copied works on machinery in the 16th century. 194 detailed plates of water pumps, grain mills, cranes, more. 608pp. 9 × 12.
25497-6 Clothbd. $34.95

LINEAR PROGRAMMING AND ECONOMIC ANALYSIS, Robert Dorfman, Paul A. Samuelson and Robert M. Solow. First comprehensive treatment of linear programming in standard economic analysis. Game theory, modern welfare economics, Leontief input-output, more. 525pp. 5⅜ × 8½. 65491-5 Pa. $12.95

ELEMENTARY DECISION THEORY, Herman Chernoff and Lincoln E. Moses. Clear introduction to statistics and statistical theory covers data processing, probability and random variables, testing hypotheses, much more. Exercises. 364pp. 5⅜ × 8½. 65218-1 Pa. $8.95

THE COMPLEAT STRATEGYST: Being a Primer on the Theory of Games of Strategy, J.D. Williams. Highly entertaining classic describes, with many illustrated examples, how to select best strategies in conflict situations. Prefaces. Appendices. 268pp. 5⅜ × 8½. 25101-2 Pa. $5.95

MATHEMATICAL METHODS OF OPERATIONS RESEARCH, Thomas L. Saaty. Classic graduate-level text covers historical background, classical methods of forming models, optimization, game theory, probability, queueing theory, much more. Exercises. Bibliography. 448pp. 5⅜ × 8¼. 65703-5 Pa. $12.95

CONSTRUCTIONS AND COMBINATORIAL PROBLEMS IN DESIGN OF EXPERIMENTS, Damaraju Raghavarao. In-depth reference work examines orthogonal Latin squares, incomplete block designs, tactical configuration, partial geometry, much more. Abundant explanations, examples. 416pp. 5⅜ × 8¼.
65685-3 Pa. $10.95

THE ABSOLUTE DIFFERENTIAL CALCULUS (CALCULUS OF TENSORS), Tullio Levi-Civita. Great 20th-century mathematician's classic work on material necessary for mathematical grasp of theory of relativity. 452pp. 5⅜ × 8½.
63401-9 Pa. $9.95

VECTOR AND TENSOR ANALYSIS WITH APPLICATIONS, A.I. Borisenko and I.E. Tarapov. Concise introduction. Worked-out problems, solutions, exercises. 257pp. 5⅜ × 8¼. 63833-2 Pa. $6.95

TENSOR CALCULUS, J.L. Synge and A. Schild. Widely used introductory text covers spaces and tensors, basic operations in Riemannian space, non-Riemannian spaces, etc. 324pp. 5⅜ × 8¼. 63612-7 Pa. $7.00

A CONCISE HISTORY OF MATHEMATICS, Dirk J. Struik. The best brief history of mathematics. Stresses origins and covers every major figure from ancient Near East to 19th century. 41 illustrations. 195pp. 5⅜ × 8½. 60255-9 Pa. $7.95

A SHORT ACCOUNT OF THE HISTORY OF MATHEMATICS, W.W. Rouse Ball. One of clearest, most authoritative surveys from the Egyptians and Phoenicians through 19th-century figures such as Grassman, Galois, Riemann. Fourth edition. 522pp. 5⅜ × 8½. 20630-0 Pa. $9.95

HISTORY OF MATHEMATICS, David E. Smith. Non-technical survey from ancient Greece and Orient to late 19th century; evolution of arithmetic, geometry, trigonometry, calculating devices, algebra, the calculus. 362 illustrations. 1,355pp. 5⅜ × 8½. 20429-4, 20430-8 Pa., Two-vol. set $21.90

THE GEOMETRY OF RENÉ DESCARTES, René Descartes. The great work founded analytical geometry. Original French text, Descartes' own diagrams, together with definitive Smith-Latham translation. 244pp. 5⅜ × 8½. 60068-8 Pa. $6.00

THE ORIGINS OF THE INFINITESIMAL CALCULUS, Margaret E. Baron. Only fully detailed and documented account of crucial discipline: origins; development by Galileo, Kepler, Cavalieri; contributions of Newton, Leibniz, more. 304pp. 5⅜ × 8½. (Available in U.S. and Canada only) 65371-4 Pa. $7.95

THE HISTORY OF THE CALCULUS AND ITS CONCEPTUAL DEVELOPMENT, Carl B. Boyer. Origins in antiquity, medieval contributions, work of Newton, Leibniz, rigorous formulation. Treatment is verbal. 346pp. 5⅜ × 8½. 60509-4 Pa. $6.95

THE THIRTEEN BOOKS OF EUCLID'S ELEMENTS, translated with introduction and commentary by Sir Thomas L. Heath. Definitive edition. Textual and linguistic notes, mathematical analysis. 2500 years of critical commentary. Not abridged. 1,414pp. 5⅜ × 8½. 60088-2, 60089-0, 60090-4 Pa., Three-vol. set $26.85

A HISTORY OF VECTOR ANALYSIS: The Evolution of the Idea of a Vectorial System, Michael J. Crowe. The first large-scale study of the history of vector analysis, now the standard on the subject. Unabridged republication of the edition published by University of Notre Dame Press, 1967, with second preface by Michael C. Crowe. Index. 278pp. 5⅜ × 8½. 64955-5 Pa. $7.00

THE HISTORICAL ROOTS OF ELEMENTARY MATHEMATICS, Lucas N.H. Bunt, Phillip S. Jones, and Jack D. Bedient. Fundamental underpinnings of modern arithmetic, algebra, geometry and number systems derived from ancient civilizations. 320pp. 5⅜ × 8½. 25563-8 Pa. $7.95

CALCULUS REFRESHER FOR TECHNICAL PEOPLE, A. Albert Klaf. Covers important aspects of integral and differential calculus via 756 questions. 566 problems, most answered. 431pp. 5⅜ × 8½. 20370-0 Pa. $7.95

CATALOG OF DOVER BOOKS

CHALLENGING MATHEMATICAL PROBLEMS WITH ELEMENTARY SOLUTIONS, A.M. Yaglom and I.M. Yaglom. Over 170 challenging problems on probability theory, combinatorial analysis, points and lines, topology, convex polygons, many other topics. Solutions. Total of 445pp. 5⅜ × 8½. Two-vol. set.
Vol. I 65536-9 Pa. $5.95
Vol. II 65537-7 Pa. $5.95

FIFTY CHALLENGING PROBLEMS IN PROBABILITY WITH SOLU-TIONS, Frederick Mosteller. Remarkable puzzlers, graded in difficulty, illustrate elementary and advanced aspects of probability. Detailed solutions. 88pp. 5⅜ × 8½.
65355-2 Pa. $3.95

EXPERIMENTS IN TOPOLOGY, Stephen Barr. Classic, lively explanation of one of the byways of mathematics. Klein bottles, Moebius strips, projective planes, map coloring, problem of the Koenigsberg bridges, much more, described with clarity and wit. 43 figures. 210pp. 5⅜ × 8½.
25933-1 Pa. $4.95

RELATIVITY IN ILLUSTRATIONS, Jacob T. Schwartz. Clear non-technical treatment makes relativity more accessible than ever before. Over 60 drawings illustrate concepts more clearly than text alone. Only high school geometry needed. Bibliography. 128pp. 6⅛ × 9¼.
25965-X Pa. $5.95

AN INTRODUCTION TO ORDINARY DIFFERENTIAL EQUATIONS, Earl A. Coddington. A thorough and systematic first course in elementary differential equations for undergraduates in mathematics and science, with many exercises and problems (with answers). Index. 304pp. 5⅜ × 8¼.
65942-9 Pa. $7.95

FOURIER SERIES AND ORTHOGONAL FUNCTIONS, Harry F. Davis. An incisive text combining theory and practical example to introduce Fourier series, orthogonal functions and applications of the Fourier method to boundary-value problems. 570 exercises. Answers and notes. 416pp. 5⅜ × 8½.
65973-9 Pa. $8.95

THE THOERY OF BRANCHING PROCESSES, Theodore E. Harris. First systematic, comprehensive treatment of branching (i.e. multiplicative) processes and their applications. Galton-Watson model, Markov branching processes, electron-photon cascade, many other topics. Rigorous proofs. Bibliography. 240pp. 5⅜ × 8½.
65952-6 Pa. $6.95

AN INTRODUCTION TO ALGEBRAIC STRUCTURES, Joseph Landin. Superb self-contained text covers "abstract algebra": sets and numbers, theory of groups, theory of rings, much more. Numerous well-chosen examples, exercises. 247pp. 5⅜ × 8½.
65940-2 Pa. $6.95

GAMES AND DECISIONS: Introduction and Critical Survey, R. Duncan Luce and Howard Raiffa. Superb non-technical introduction to game theory, primarily applied to social sciences. Utility theory, zero-sum games, n-person games, decision-making, much more. Bibliography. 509pp. 5⅜ × 8½. 65943-7 Pa. $10.95

Prices subject to change without notice.
Available at your book dealer or write for free Mathematics and Science Catalog to Dept. GI, Dover Publications, Inc., 31 East 2nd St., Mineola, N.Y. 11501. Dover publishes more than 175 books each year on science, elementary and advanced mathematics, biology, music, art, literary history, social sciences and other areas.